45

Advances in Polymer Science

Fortschritte der Hochpolymeren-Forschung

W0037128

Interactions Between Macromolecules in Solution and Intermacromolecular Complexes

By E. Tsuchida and K. Abe

With 52 Figures

Springer-Verlag
Berlin Heidelberg GmbH 1982

ISBN 978-3-662-15754-1   ISBN 978-3-540-39443-3 (eBook)
DOI 10.1007/978-3-540-39443-3

Library of Congress Catalog Card Number 61-642

© Springer-Verlag Berlin Heidelberg 1982
Originally published by Springer-Verlag Berlin Heidelberg New York in 1982
Softcover reprint of the hardcover 1st edition 1982

2152/3140 – 543210

Table of Contents

Interactions Between Macromolecules in Solution and Intermacromolecular Complexes

Eishun Tsuchida and Koji Abe[*]

Department of Polymer Chemistry, Waseda University, Tokyo 160, Japan

Macromolecules with complementary binding sites associate almost stoichiometrically in solution to form the "intermacromolecular complex". Studies on intermacromolecular complexes are mainly concerned with the elucidation of various reactions in vivo essential for the maintenance of life. They also provide indispensable information on the production of new macromolecular materials.

In this review, the results of fundamental studies on the characteristics of intermacromolecular complexes considering their formation, structure and physical and chemical properties are systematically discussed. These complexes are divided into four classes on the basis of their main interaction forces, i.e. polyelectrolyte complexes, hydrogen-bonding complexes, stereocomplexes and charge-transfer complexes. The cooperative phenomena of the formation of intermacromolecular complexes are also discussed quantitatively. Moreover, up-to-date topics from various fields and their various possibilities of application for future developments are mentioned.

* Present address:
  Department Functional Polymer Science, Shinshu University, Ueda 386, Japan

Advances in Polymer Science 45
© Springer-Verlag Berlin Heidelberg 1982

1 Introduction

The characteristics of macromolecules are not only governed by the chemical structure of the polymer chain but also by the formation of macromolecular aggregates. That is, macromolecules are characterized not only by *first-order structures* but also by *higher-order structures* such as configuration and conformation of polymer chains. Furthermore, macromolecules may aggregate due to *secondary binding forces,* solvation, steric factors, and interpenetration. As reflected by the reaction mechanisms *in vivo,* specific interactions among inter- and intramacromolecules are necessary for the display of a functionality regardless of the polymer chain structure. However, in spite of the importance of these interactions, relatively few studies on assemblies formed through interactions between synthetic polymer chains have been reported. Since synthetic polymers have simpler structures than biopolymers, the functional phenomena occurring in complicated reactions in living cells may be more easily understood. Therefore, suggestions about the design of the functions on the polymer chains are given to study the mechanism of complexation between synthetic macromolecules.

This article summarizes formation, properties and structure of intermacromolecular complexes between synthetic macromolecules mainly and outlines their applications.

2 General Features of Intermacromolecular Complexes

2.1 Macromolecular Associations in Solution

Macromolecules aggregate with each other in solution. These aggregation phenomena are ordinarily observed as phase separation phenomena, such as precipitation, gelation, coacervation, and emulsion, or crystallization and liquid-crystallization of polymers or self-assembly of subunits of biopolymers. The regularities of orientation of macromolecular chains are changed from random mixing states to specific higher-order structure according to the above mentioned order, and the terms "aggregate", "associate" and "assembly" designate such changes in the orientation from random to specifically ordered states.

In dilute solution, all macromolecular chains undergo interactions with each other resulting in the so-called intermolecular excluded volume effect, corresponding to the intermolecular potential. This effect is also observed if one does not assume particular cohesive forces to occur between the macromolecular chains. Under these conditions, the second virial coefficient is calculated from the equation[1, 2];

$$A_2 = \frac{N_A \beta_0}{2 M^2} = \frac{N_A}{2} \left(\frac{n}{M}\right)^2 \cdot \beta_0^s \cdot h(\overline{Z})$$

$$= 4\pi^{3/2} N_A (\langle S^2 \rangle^{3/2}/M^2) \overline{Z} \cdot h(\overline{Z}) \tag{1}$$

where N_A = Avogadro's number, n = the number of segments, β_0^s = excluded volume between segments, M = moles of solute, $Z = (3/2 \pi a^2)^{3/2} \beta_0^s n^{1/2}$, $\langle S^2 \rangle$ = mean square radius of gyration. Therefore, only in the ideal state, i.e. in the Θ state, can interactions between macromolecular chains be ignored. In contrast, in concentrated systems, macromolecular chains interpenetrate, and this results in a slight change of the intermolecular potential with varying distance between macromolecules. Thus, the intermacromolecular excluded volume decreases, due to the formation of a random mixing state of macromolecules. Under such conditions, the excluded volume β is defined as follows;

$$\beta = \frac{\bar{v}_1}{\phi_2} \left[\frac{1}{\phi_1^{-1} - (1 - m^{-1}) - 2\chi\phi_2} + m \right] \tag{2}$$

where $m = \overline{V}_2/\overline{V}_1$, V = partial molar volume, ϕ = volume fraction, $\chi = (1 - N_A\beta_0^s/\overline{V}_1)/2$. In appendices 1 and 2 are listed the solvents and macromolecules, respectively.

In concentrated polymer solutions, interpenetration phenomena of polymer chains and their structures were studied by Stutz et al.[3]. They suggested that the interpenetration of polymer chains was less than 16–18% in polymer-polymer crosslinking systems and that the polymer chains were contracted. It may be assumed that interpolymer interactions between pairs of macromolecules with low compatibility relative to each other occur at the surface of the polymer domain in a sphere model. However, in the case of complementary polymers, e.g. pairs of polymers displaying the ability of forming complexes through comparatively strong secondary binding forces (polycation-polyanion, polymers forming hydrogen bonds between specific active sites, etc.), the polymer chains can intertwine more efficiently. Under the most suitable conditions, a ladder-like structure may be formed, e.g. a double-stranded helix structure of DNA[4] and a stereocomplex[5].

The quantity of β_0^s is estimated by the free energy change of mixing (ΔG). Whether macromolecules are separated from the solution or not, i.e. whether macromolecules are compatible or not, is indicated by the sign (negative or positive) of ΔG. In the case of $\Delta H > 0$ and $\Delta S > 0$, phase change of the macromolecular solution from imcompatible to compatible is observed with rising temperature (there exists an upper critical solution temperature; UCST), while in the case of $\Delta H < 0$ and $\Delta S < 0$, phase change from compatible to incompatible polymer solution occurs (there exists a lower critical solution temperature; LCST). In general, in concentrated polymer solutions, when increasing the temperature, the solution tends to undergo the following changes: homogeneous → heterogeneous → homogeneous. Flory[6], Prigonine[7] and Patterson[8] explained this phenomenon by the free volume theory: in addition to coordination entropy arising through mixing and peculiar intermolecular interactions in corresponding states, they introduced the theory of the change of state by mixing (mainly change of volume). These phase separation phenomena have been observed in the systems of poly(vinyl alcohol)[9, 10], poly(ethylene oxide)[11], copolymers of styrene and maleic anhydride[12], collagen[13], polyelectrolyte[14], polystyrene[15], poly(γ-benzyl-L-glutamate)[16], agarose[17], etc.

Since the compatibility of macromolecules is extensively discussed in many works on the χ-parameter[18], it is treated here only briefly. When considering the compatibility of polymers, one must naturally give thought to weak intermacromolecular interactions, for example, van der Waals force, dipole-dipole interaction, and so on.

In these aggregation phenomena, the formation of intermacromolecular complexes is attributed to the fact that macromolecules with complementary binding sites interact with each other almost stoichiometrically in solution due to certain secondary binding forces. In this review, the formation of complementary complexes in synthetic macromolecular systems is mainly discussed.

2.2 Secondary Binding Forces Between Intermacromolecular Complexes

Intermolecular interactions were introduced for the first time by van der Waals in 1873; he thus attempted to explain the deviation of the real gas from the ideal gas. In order to apply the ideal gas law equation to the behavior of real gases, allowance should be made for the attractive and repulsive forces occurring between molecules. From that time on, the dipole moment theory of Debye (1912) and the dispersion energy or induced dipole theory by London (1930) were the main driving forces of the research about intermolecular interactions.

When two different molecules (A and B) approach, the molecular energy is changed by the following phenomena
1) overlap of electron clouds of A and B
2) exchange of electrons
3) electron transfer or delocalization of localized electrons
4) changes of the electronic states of A and B (the polarization may be due to the change in the electron distribution and/or mixing of excited states)
5) electrostatic interactions (between atomic nucleus of A (B) and electron of B (A), between two electrons (of A and B) and between two atomic nucleus (of A and B)
6) dipole-dipole interactions when A and B have dipole moments. These factors never act individually in real cases; i.e. some factors are observed at the same time cooperatively or concertedly, or one phenomenon induces another one. These factors must therefore be discussed simultaneously.

Secondary binding forces are mainly classified into Coulomb forces, hydrogen-bonding forces, van der Waals forces, charge transfer forces, exchange repulsion and hydrophobic interactions (Table 1). Besides these forces, there are other interactions such as ion-dipole and solvophobic interactions.

2.2.1 Coulomb Forces (Electrostatic Interaction Forces)

The Coulomb force acts between two charged molecules, for example, $Na^+ \ldots \ldots Cl^-$. According to the point charge model, the energy is described by the Coulomb law.

$$E_{es} = Z_A \cdot Z_B \cdot e^2/R \tag{3}$$

(Z_A and Z_B are valencies of charges of A and B and R is the distance between A and B)
This interaction is characterized by comparatively long-range and relatively strong forces, being about several tens of kcal/mol and differing from other interaction forces.

Table 1. Characteristics of secondary binding forces

Classification	Molecules	Interaction energy	Notes
Coulombic force	Ion–Ion	$U = Z_A \cdot Z_B \cdot e^2/R$	Attraction and repulsion Long-range force
Hydrogen bond	Proton acceptor – Proton donor	$\begin{cases}\text{Stockmayer equation}\\ \text{Lippincott-Schröder equation}\\ \text{Scheraga equation}\end{cases}$	Existence of direction for bonding Attraction force
van der Waals force			
Orientational force	Permanent dipole – Permanent dipole	$U = -\dfrac{2}{3} \cdot \dfrac{\mu_2^2 \mu_1^2}{R^6} \cdot \dfrac{1}{kT}$	Attraction force
Induced force	Permanent dipole – Induced dipole	$U = -\dfrac{(\alpha_2 \mu_1^2 + \alpha_1 \mu_2^2)}{R^6}$	
Dispersion force	Transient dipole – Induced dipole	$U = -\dfrac{2}{3} \cdot \dfrac{I_1 I_2}{I_1 + I_2} \cdot \dfrac{\alpha_1 \alpha_2}{R^6}$	Short-range force
Charge-transfer interaction	Electron acceptor – Electron donor	$U \simeq -\dfrac{\beta^2}{I_b - A_a + C}$	Attraction force
Exchange repulsion	All molecules	$U = A \cdot \exp(-BR)$	Very short-range repulsion
Hydrophobic interaction	Hydrocarbon molecules in water	$\begin{cases}\Delta F = 2.436 + 0.884\,n\ (\text{Alkane})\\ \Delta F = 1.503 + 0.884\,n\ (\text{Alkene})\\ \Delta F = 0.903 + 0.860\,n\,(\text{Dialkane})\end{cases}$	Force caused by the specific structures of water molecules

In polyelectrolyte systems, the theory is adjusted either to the point charge model assuming a distribution of point charges on the polymer chain or to the dipole-ion theory considering an ion pair as a dipole. Their potential energies are expressed as

$$U_{ij} = q_i q_j / D r_{ij} \tag{4}$$

$$U_{ij} = [(\mu_i \cdot \mu_j) - 3(\mu_i \cdot r_{ij})(\mu_j \cdot r_{ij})/r_{ij}^2]/(Dr_{ij}^3) \tag{5}$$

2.2.2 Hydrogen Bonds

Hydrogen bonds arise through interactions between electron-deficient hydrogen atoms and atoms of high electron density. There are two main types of hydrogen bonds
(1) hydrogen bonds which connect atoms with an electronegativity higher than that of hydrogen, e.g. H_2O . . . . . H–OH
(2) hydrogen bonds which connect atoms of lower electronegativity such as B–H–B bonds in boranes.

The potential energy of the hydrogen bond is explained by the Stockmayer equation[19] (based on the electrostatic potential), the Lippincott-Schröder equation[20] (based on chemical bonds) and the Scheraga equation[21] (based on van der Waals and Coulomb interactions). The most remarkable features of hydrogen bonds are
(1) the hydrogen bond energy is comparatively low, 3–6 kcal/mol
(2) there is a pronounced directionality on bonding
(3) the hydrogen bond is expressed by the following equilibrium: X–H . . . Y ⇋ X⁻ . . . H⁺–Y.

2.2.3 Van der Waals Forces

Van der Waals forces are relatively short-range forces between molecules with permanent dipoles or molecules with induced dipoles, i.e. they almost occur between all molecules. These interactions include dipole-dipole interactions, dipole-induced dipole interactions and dispersion energy.

Orientational energy is due to dipole-dipole interactions. If molecules with a permanent dipole assume a random arrangement, the average potential energy should be zero. Generally, however, the probability of taking an arrangement of low energy is high resulting in attraction forces. In the case of two polar molecules with constant distance R, the potential function is calculated according to the equation in Table 1, assuming a Boltzmann distribution for the disposition of the four charged molecular centers according to the Coulomb law. Induced forces act between a polar molecule with a permanent dipole and a neutral molecule with an induced dipole due to the electric field of the permanent dipole (see Table 1). Dispersion energy is observed with certain molecules[22] which is explained as follows. Even in completely symmetrical molecules, for example, He and Ar, the electron distribution immediately becomes asymmetric so that an instantaneous dipole moment (transient dipole) results. Such a transient dipole induces dipoles in other molecules leading to interactions between the latter (for the interaction energy see Table 1). Van der Waals forces are at most about one kcal/mol and relatively short-range interaction forces.

2.2.4 Charge-Transfer Interactions

Charge-transfer interactions are attractive forces caused by charge-transfer between an electron donor (with low ionic potential) and an electron acceptor (with high electron affinity)[23]. Therefore, the potential energy is expressed as shown in Table 1, where I_D, A_A, β and C denote the ionic potential of the electron donor, the electron affinity of the electron acceptor, the electron exchange energy, and a constant, respectively.

2.2.5 Exchange Repulsion

When two molecules come so close to each other that both electron clouds can overlap, electron exchange takes place. This gives rise to repulsive forces between molecules.

2.2.6 Hydrophobic Interactions

Hydrophobic interactions occur when hydrocarbons are dissolved in aqueous medium. They essentially differ from the interaction forces mentioned previously because they are not caused by direct cohesive forces between molecules but by the specific structure of water molecules. A water molecule forms four hydrogen bonds with neighboring water molecules to form a cluster. Kauzmann[24] pointed out the effect of the aggregation of bulky and non-polar (hydrophobic) groups on the stabilization of the higher order structure of proteins (hydrophobic interaction). When low molecular weight hydrocarbons are transferred from a non-polar solvent to water, the changes of the thermodynamic parameters are $\Delta S < 0$, $\Delta H < 0$, and $\Delta F > 0$. Thus, the low stability of hydrocarbons in water is due to the large decrease of entropy. This phenomenon was described qualitatively by Frank and Evans[25] assuming that the water molecules surrounding the non-polar solute is more highly orientated (ice-berg). To sum up, in order to minimize the contact surface area of hydrophobic groups with water, resulting in a decrease of the entropy, hydrophobic groups aggregate with each other. Nemethy and Scheraga[26] assumed a flickering-cluster model for the water molecule and reported that the hydrophobic interactions are governed by entropy. Tanford[27] as well as Nemethy and Scheraga compared the theoretical and experimental values. Factors determining hydrophobic interactions are, in addition to the change of the structure of water mentioned above, van der Waals forces between hydrophobic groups and the restraint of the internal degree of freedom (mainly rotation) of hydrophobic groups caused by clustering of water molecules.

Similarly, attractive forces called „solvophobic interaction"[28] exist when the interaction forces between solvent molecules are stronger than those between solvent and solute molecules.

Even in macromolecular systems, the secondary binding forces mentioned up to here act in the same manner as in the case of low molecular weight compounds, if one considers secondary bonds individually. However, for all practical purposes, they act at the same time in an extremely complicated manner, concertedly and never separately. Moreover, each active site of the molecule interacts cooperatively with other sites because of the neighboring effect (for details see Sect. 4). Therefore, in intra- and intermacromolecular interaction systems, it is quite difficult to investigate separately the effective secondary binding forces, and it should be noted that the total interaction force might not be the sum of the individual binding forces.

2.3 Importance of Intermacromolecular Interactions in Biological Systems

Many molecules including biopolymers participate in biological functions as a molecular assembly or tissue: the self-assembly of the microtublin of bacterial flagella, antigen-antibody reactions, the high activity and selectivity of enzymes, etc. are skillfully and accurately achieved by intermacromolecular interactions.

2.3.1 Self-Assemblies of Proteins

Many enzymes, virus shells, bacterial pili, microtubles, and other biological structures are assemblies of protein subunits; most typical examples are tobacco mosaic virus[29-32], phase change of proteins[33], polymerization of actin and myosin[34, 35], and bacterial flagella of *Salmonella*[36]. Such mechanisms of self-assembly suggest that the interactions between macromolecules and their combination are specific processes, which implies that the interaction forces in macromolecular systems are much stronger than those in low molecular weight systems.

2.3.2 Molecular Recognition

Molecular recognition is one of the most important phenomena in biological systems such as antigen-antibody and enzyme reactions and the mechanism of the memory and transmission of genetic information. Three main theories of molecular recognition in antigen-antibody and enzymic reactions have been established: the *key-key lock theory*[37], the *induced-fit theory*[38] and the *rack model*[39].

In order to store, transfer and transmit the genetic information with extreme accuracy, molecular recognition between specific base pairing of nucleic acids through hydrogen bonding is essential. Allosteric effects typically observed in the oxygenation of hemoglobin which is composed of four protein subunits, are realized through continuous transmission of such an information as the conformational change of the macromolecular chain.

2.3.3 Biomembranes

With the development of the study about life science, particular interest has been focused on the elucidation of life phenomena. The minimum unit of a living body is a cell. The cell membrane represents a specific molecular associate of certain molecules such as lipids and proteins molecules. Its structure is considered to consist of macromolecules (proteins in this case) floating in vesicular bilayer membranes, termed "fluid mosaic model"[40]. In order to clarify the mechanism and the general laws of molecular association, it is important to study more intensively the cell membranes as molecular associates. Various functions of biomembranes, e.g. transport, selective permeability of micro ions and specific reactions on the membrane surface, may be caused by the fact that cell membranes themselves are the molecular associates (especially the existence of proteins in the membrane in contrast to model membranes such as vesicular bilayers). Membrane proteins play an important role in the determination of the overall structure and some functions of the cell membranes. It may be assumed that these proteins are incorporated into lipid bilayer membranes through hydrophobic interaction forces. There are some interactions (e.g. motional behavior and distribution) between lipids and proteins. Interactions between lipids and membrane proteins have been detected as the changes of molecular motions of lipids by means of NMR[41–43]. The interaction of synthetic polymers with vesicular membranes has also been studied as model systems to clarify the protein-lipid interaction. Up to now, interactions between synthetic polypeptides and phospholipid bilayers have mainly been studied[44–47]. Hammes and Schullery observed pronounced interactions to occur between poly(L-lysine) and phosphatidylserine[46]. For details see Chan et al.[45].

It may be said that in biological systems specific interactions between macromolecules are quite important and that the various functions realized in these systems are closely related to higher-order structures and assemblies of macromolecules controlled by intra- and intermacromolecular interactions. Therefore, it is necessary to discuss collectively the interrelationship between interaction forces, the structure of polymers and the functionalities of macromolecules as shown in Fig. 1.

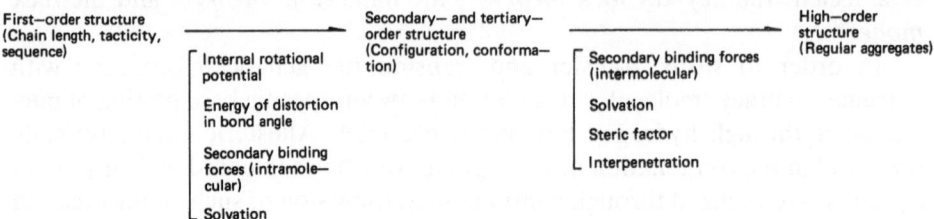

Fig. 1. Controlling factors in the formation of macromolecular aggregates

3 Formation, Structure and Properties of Intermacromolecular Complexes

Polymeric structures within which the primary macromolecular units are held together by ionic linkages have received relatively little attention by the material specialists. However, ionically bonded organic polymeric structures have aroused the interest of biochemist and biophysicists, due to the fact that the majority of biopolymers are ionically associated in living systems. The extraordinary complexity and instability of biological systems seem to have discouraged the applied polymer scientists and engineers from considering synthetic polymeric materials crosslinked through secondary binding forces as products of practical and commercial utility.

At the end of the 19th century, Kossel[389] reported the phase separation: there are the polymer condensed phase and the polymer diluted phase in the aqueous solution systems of proteins and carbohydrates. On the contrary, Bungenberg de Jong[595] reported the liquid-liquid phase separation in the gelatin-arbic gum system. Such phase separation phenomenon is called "complex coacervation", and the polymer condensed liquid phase is called "complex coacervate". Oparin investigated that such complex coacervates are important in the origin of life (prebiological systems). Thereafter, complex coacervates are watched and obtained by various pairs of oppositely charged polyelectrolytes. In 1949 polyelectrolyte complexes composed of synthetic polyelectrolytes were first reported by Fuoss[80], in poly(vinyl-N-butylpyridinium bromide)-poly(sodium styrenesulfonate) system. However, polyelectrolyte complexes have been actively studied since Michaels et al. reported formation, properties, and possibility of various applications of polyelectrolyte complexes[49]. Since then different kinds of intermacromolecular complexes are focused and studied energetically.

Macromolecular chains, as stated in the previous chapter, may undergo interactions in solution except in the ideal state (Θ state). In this chapter, the association phenomena of more than two different macromolecular chains in solution caused by secondary binding forces such as electrostatic interactions are discussed. The obtained associates are generally called "intermacromolecular (interpolymer) complexes" or „polymer-polymer complexes".

3.1 Classification and General Characteristics of Intermacromolecular Complexes

Intermacromolecular complexes are divided into the following four classes on the basis of the main interaction forces. However, it should be noted that this classification is only made for simplicity.

(A) Polyelectrolyte complexes

$$\text{polyanion} + \text{polycation} \xrightarrow[\text{Coulomb forces}]{} \text{complex} + \text{microion} \qquad (6)$$

$$(7)$$

$$(8)$$

(B) Hydrogen-bonding complexes

$$\underset{\text{polymer}}{\text{proton-accepting}} + \underset{\text{polymer}}{\text{proton-donating}} \xrightarrow[\text{hydrogen bond}]{} \text{complex} \qquad (9)$$

$$(10)$$

$$(11)$$

(C) Sterocomplexes

iso-poly(methyl methacrylate) + synd-poly(methyl methacrylate)

$$\xrightarrow[\text{van der Waals force}]{} \text{complex} \qquad (12)$$

$$(13)$$

o CH$_3$
o O
• C

(D) Charge transfer complexes

$$\underset{\text{polymer}}{\text{electron-accepting}} + \underset{\text{polymer}}{\text{electron-donating}} \xrightarrow[\substack{\text{interactions}}]{\text{charge-transfer}} \text{complex} \tag{14}$$

$$\cdots\text{O(CH}_2)_2\text{NH(CH}_2)_2\text{OCOO} \left\langle\text{benzene}\right\rangle \underset{\overset{|}{\text{Me}}}{\overset{\text{Me}}{\underset{|}{\text{C}}}} \left\langle\text{benzene}\right\rangle \text{OCO} \cdots \quad + \tag{15}$$

$$\cdots\text{O(CH}_2)_2\text{OCO}\left\langle\text{benzene-NO}_2\right\rangle\text{COO(CH}_2)_2\text{OCOO}\left\langle\text{benzene}\right\rangle \underset{\overset{|}{\text{Me}}}{\overset{\text{Me}}{\underset{|}{\text{C}}}} \left\langle\text{benzene}\right\rangle \text{OCO}\cdots$$

Polyelectrolyte complexes are formed by mixing oppositely charged polyelectrolytes, i.e. polyanions and polycations, due to Coulomb forces. Simultaneously, microions are released almost quantitatively. Complexes containing hydrogen bonds (hydrogen-bonding complexes) are formed by combination of polymers bearing proton-accepting units and proton-donating units. Stereocomplexes are generated by combination of isotactic with syndiotactic poly(methyl methacrylate) (PMMA), mainly through van der Waals forces. Stereocomplexes exhibit a specific higher-order structure, e.g. syndiotactic PMMA with a β-pleated sheet structure intertwined in the grooves of the helical structure of isotactic PMMA. Charge transfer complexes are formed in systems of electron-accepting polymers and electron-donating polymers through charge-transfer interactions. The general characteristics of these complexes are summerized in Table 2.

Polyelectrolyte complexes are divided into four subclasses by a combination of strong and weak polyelectrolytes. These polyelectrolyte complexes are denoted and analyzed by their eletrochemical behavior, owing to the released microions, the hydrodynamic properties, the changes of molecular weight, the conformation and radius of gyration, and the physical and chemical properties of the products. In general, the composition of polyelectrolyte complexes depends on the degree of dissociation of the polyelectrolyte components. Therefore, in the system of strong polybase-strong polyacid, the composition of the obtained complexes is unity. On the other hand, the composition markedly depends on the degree of dissociation when weak polyelectrolytes are used.

The formation of complexes containing hydrogen bonds is detected by the same methods as in the case of polyelectrolyte complexes. Moreover, spectroscopic methods such as Infrared, Raman and Nuclear Magnetic Resonance are used as efficient analytical methods. The composition of [proton-donating polymer unit]/[proton-accepting polymer unit] is generally 2/1 to 2/3 in dilute

Table 2. Classification and general characteristics of interpolymer complexes

Interpolymer complex	Analytical method	Composition	Solubility
Polyelectrolyte complex			
Strong polyacid-Strong polybase	Conductivity, Potentiometry, Electrophoresis, Viscosity,	1/1	Specific ternary solvent
Strong polyacid-Weak polybase	Sedimentation, Diffusion, Centrifugation, Light	1/1 ~ 1/5	High pH, High μ[a]
Weak polyacid-Strong polybase	scattering, Turbidity, IR spectroscopy, Elemental	9/1 ~ 1/1	Low pH, High μ
Weak polyacid-Weak polybase	analysis, X-ray diffraction, Optical and electron microscopy, etc.,	Various	High and low pH, High μ
Hydrogen-bonding complex			
Poly(carboxylic acid)-Polyether	IR, NMR and Raman spectroscopy, Hydrodynamic	1/1[b], 2/3[c]	High pH, DMSO[d]
Poly(carboxylic acid)-PVPo[e]	methods mentioned above, Electrochemical methods	3/2 ~ 1/1	High pH, DMSO
Poly(carboxylic acid)-PVA[f]	mentioned above, Detection of physical and chemical	1/1	High pH, DMSO
Poly(carboxylic acid)--P=O containing polymer	properties of the products mentioned above, Thermal analysis, etc.,	1/1	HMPA[g]
Stereocomplex			
In polar solvent	Hydrodynamic methods mentioned above, Detection of	1/2 (precipitate)	Chloroform,
In nonpolar aromatic solvent	physical and chemical properties of the products, X-ray diffraction, Thermal analysis, Absorption spectra, etc.,	1/2 (gel)	CH$_2$Cl$_2$
Charge-transfer complex	Visible and ultraviolet spectroscopy	1/1	

[a] Ionic strength, [b] In dilute solution system, [c] In highly concentrated system, [d] Dimethyl sulfoxide, [e] Poly(N-vinyl-2-pyrrolidone), [f] Poly(vinyl alcohol), [g] Hexamethylphosphoric-triamide

solution systems and 2/3 in concentrated solution systems. This kind of complex is soluble in very strong proton-accepting aprotic solvents.

The analytical methods of the determination of stereocomplexes are the same as those mentioned previously, especially the utilization of the hydrodynamic properties of the complex solution and the physicochemical properties of the products. In polar organic solvents, the complex is obtained as precipitate and its composition is 1/2 according to the molar ratio of the units [iso-PMMA]/[synd-PMMA]. In contrast, in non-polar aromatic solvents, the com-

Table 3. Formation of intermacromolecular complexes between synthetic macromolecules

Polymer A	Polymer B	Ref.
Polyelectrolyte Complexes		
Poly(sodium styrenesulfonate)	Poly(4-vinylbenzyltrimethylammonium chloride)	48–77
Poly(sodium styrenesulfonate)	Quaternized poly(4-vinyl-pyridine)	78–80
Poly(sodium styrenesulfonate)	Poly[1-(4-[2-(triethylammonio)ethyl]-phenyl)ethylene bromide]	81–85
Poly(sodium styrenesulfonate)	Integral-type polycations	86–88
Poly(sodium styrenesulfonate)	Poly(2-N,N-dimethylaminoethylmeth-acrylate)	89, 90
Poly(sodium styrenesulfonate)	Poly(1,2-dimethyl-5-pyridinium methyl sulfate)	91
Poly(sodium styrenesulfonate)	Polyviologen	92–94
Poly(4-vinylbenzylaminedinitro-benzoyloxyphenol)	Poly(dimethylvinylbezylamine)	95
Poly(sodium vinylsulfonate)	Integral-type polycations	96
Poly(sodium vinylsulfonate)	Poly(ethylenimine)	97
Poly(carboxylic acid)s	Poly(ethylenimine)	98–114
Poly(carboxylic acid)s	Poly(ethylenepiperazine)	115
Poly(carboxylic acid)s	Poly(4-vinylpyridine)	116–120
Poly(carboxylic acid)s	Integral-type polycations	121–141
Poly(carboxylic acid)s	Poly(4-vinylbenzyl-trimethylammonium chloride)	142, 143
Poly(carboxylic acid)s	Quaternized poly(4-(or 2-)vinyl pyridine)	144-149
Poly(carboxylic acid)s	Poly(2-N,N-dimethylaminoethyl meth-acrylate)	115, 150
Poly(carboxylic acid)s	Poly(vinylaminoacetal)	151
Carboxymethylated poly(vinyl alcohol)	Aminoacetalyzed poly(vinyl alcohol)	152–157
Sulfated poly(vinyl alcohol)	2,2-Diethoxyethyl-trimethyl-ammonium of poly(vinyl alcohol)	158–164
Sulfated poly(vinyl alcohol)	Poly(ethylenimine)	165–169
Polyampholites	Polyanions and polycations	170
Polyphosphate	Polybase	171–174
Intermacromolecular Complexes Formed by Hydrogen Bonds		
Poly(carboxylic acid)s	poly(ethylene oxide)	175–221
Poly(carboxylic acid)s	Poly(N-vinyl-2-pyrrolidone)	222–261
Poly(carboxylic acid)s	Poly(vinyl alcohol)	262–283
Poly(carboxylic acid)s	Poly(acrylamide)	284, 285
Copolymers containing carboxylic acid units	Copolymers containing 4-vinylpyridyl units	286, 287
Poly(carboxylic acid)s	Poly(1,2-dimethoxyethylene)	288–290
Poly(carboxylic acid)s	Poly(dimethyltetramethylene-phosphoric triamide)	291–293

Table 3 (continued)

Polymer A	Polymer B	Ref.
Poly(carboxylic acid)s	Poly(vinylmethyl ether)	294
Poly(carboxylic acid)s	Poly(vinylbenzo-18-crown-6)	295
Poly(vinyl alcohol)	Poly(N-vinyl-2-pyrrolidone)	296
Poly(vinyl alcohol)	Poly(acrylamide)	297
Polyarylate	Poly(ethyelene oxide)	298
Intermacromolecular Complexes Through van der Waals Forces		
Iso-poly(methyl methacrylate)	Synd-poly(methyl methacrylate)	299–374
Iso-Poly(methyl methacrylate)	Synd-poly(methacrylic acid)	375, 376
Poly(vinyl chloride)	Synd-poly(methyl methacrylate)	377–379
Polystyryl lithium	Synd-poly(methyl methacrylate)	380
Charge-Transfer Complexes		
Electron donating polymers	Electron accepting polymers	381–386 592

plex is obtained as a gel but its composition is also 1/2. The complexes are soluble in $CHCl_3$ and CH_2Cl_2.

Charge-transfer complexes are mainly characterized by their ultraviolet and visible spectra. The ratio of their repeating units is almost unity.

Only few theoretical studies on the formation of such intermacromolecular complexes have been performed up to now. This may be due to the following facts

(1) the contribution of each secondary binding force cannot be clearly separated

(2) synthetic polymers essentially exhibit random molecular weights, tacticities, sequences, and secondary and tertiary structures

(3) biopolymers contain too many kinds of functional groups so that characterization of individual bonds is not possible.

Table 3 summarizes typical intermacromolecular complexes composed of synthetic macromolecules. A review of the formation of intermacromolecular complexes composed of copolymers has been published by Bekturov et al.[387].

3.2 Polyelectrolyte Complexes

Oppositely charged polyelectrolytes interact with each other to form polyelectrolyte complexes in solution, the possible combinations including strong polyacids-strong polybases, strong polyacids-weak polybases, weak polyacids-strong polybases, weak polyacids-weak polybases, or polyampholytes. Consid-

ering that almost all biopolymers are polyelectrolytes, studies on polyelectrolyte complexes of synthetic macromolecules as models of complicated biological systems are very important. Moreover, electrostatic interaction forces are
much stronger than other secondary binding forces so that the obtained products are expected to display specific physico-chemical properties. The structure
of polyelectrolyte complexes differs from that of other ion-containing macromolecules such as ionomers, ion-exchange resins and snake-cage resins[48]
(Fig. 2). Ionomers and ion-exchange resins contain only one type of polyelectrolyte which is crosslinked by divalent counterions and covalent bonds,
respectively. Snake-cage resins[388] contain two types of polyelectrolytes, but
they are too hard to be treated as industrial materials.

Studies on the interaction between oppositely charged polyelectrolytes
date back to 1896 when Kossel[389] precipitated egg albumin with protamine.
Since that time extensive studies have been made on pairs of strong polyelectrolytes, pairs of strong and weak polyelectrolytes, pairs of weak polyelectrolytes, as well as on amphoteric complexes. However, the theoretical considerations of intermacromolecular interactions between polyelectrolytes were only
based on extremely simplified model systems. However, even in the case of
such systems, there are many unsolved problems such as the determination of
the local dielectric constant in domains of macromolecular chains, the evalua-
tion of other secondary binding forces, especially hydrophobic interactions,
and so on.

Oosawa[390] described a very simple theory based on the electric free energy
to claculate the repulsive forces between parallel rod-like macoions in solution
as a function of the charge density on the rods. The total extensive force

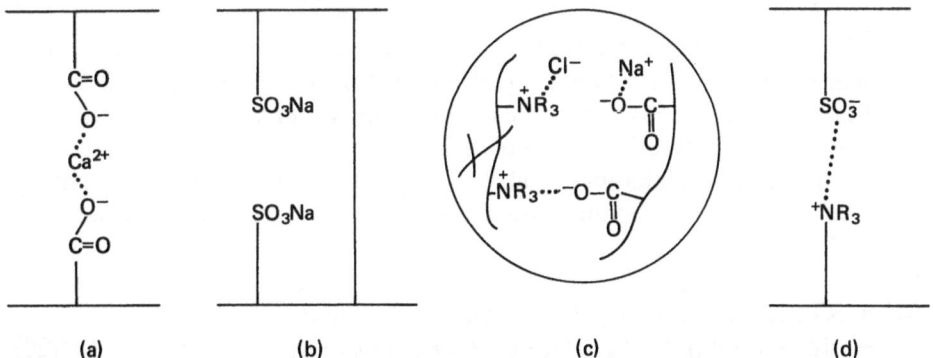

Fig. 2 a–d. Ion-containing polymer materials. (a) Ionomers: crosslinked by multivalent metal
ions (e.g. Ca^{2+} and Mg^{2+}), readily thermoformable, hydrophobic, quite insoluble, forming no gel;
(b) Ion-exchange resin: permanently crosslinked by covalent bonds, completely insoluble, not
thermoformable, volume of gels may vary with pH, having anionic or cationic ion-exchange
ability; (c) Snake-cage resin: crosslinked by both ionic and covalent bonds, reversible ion-
exchange resin with both anionic and cationic ion-exchange ability; (d) Polyelectrolyte complex:
crosslinked by ionic bonds, soluble in certain solvents, relatively hydrophilic, volume-stable gel,
thermoformable (when plasticized)

$(\partial f/\partial X)$ of an assembly of m rods with length ℓ and charge number n (charge density $= -ne_0/\ell$) at small extension X in the absence of low molecular weight salts is given by

$$X(\partial f/\partial X) = -n \cdot m(m-1) \cdot Q \cdot k \cdot T \qquad (0 \leq Q \leq 1/mz) \qquad (16)$$
$$= -n \cdot m(2/z - Q - 1/mz^2 Q) \cdot k \cdot T \qquad (1/mz \leq Q \leq 1/z) \qquad (17)$$
$$= -n \cdot m(m-1)/mz^2 Q \cdot k \cdot T \qquad (1/z \leq Q) \qquad (18)$$

where z is the number of charges of the counterions and Q $(= n \cdot e_0^2 \cdot \varepsilon \cdot k \cdot T \cdot \ell)$ is a dimensionless quantity representing the charge density. Repulsion between two parallel rods results for m = 2. At high charge density, the repulsion is much smaller than the direct Coulomb forces between charged rods, even at a short distance. The addition of low molecular weight salts does not appreciably reduce repulsion as long as the average concentration of salt ions is much lower than the concentration of counterions accumulated in the space between the rods. On the other hand, Ise[391] introduced the intra-macroion interaction free energy F_{int} using a rigid metallic sphere model for each ionized group and fixed counterion, both of which are constrained in the field of potential ψ_2. Then

$$F_{int} = L_p \sum_i \int_0^R \int_0^1 \frac{4 \cdot \pi \cdot r^2 \cdot \nu_i \cdot \psi_2'(\xi)}{v_p} e_i \cdot d\xi \cdot dr \qquad (19)$$

where ν_i denotes the number of ions of species i in a macroion, e_i its charge, $\psi'_2(\xi)$ the potential at the position of e_i (excluding the self-potential of e_i), L_p the total number of macroions in the system, and ξ the fraction of their final charges which they have at any stage of the integration, R the radius of the macroion, v_p the volume of the macroion $(= 4\pi R^3/3)$; the summation is performed over ionized groups and counterions.

Moreover, M. J. Voorn[392–397], A. Veis[398–401], A. Nakajima[153] and others[402–405] described the complex coacervation (one type of polyelectrolyte complex in the form of a liquid-liquid phase; for details see Sect. 3.2.1) as a simplified system composed of three components, water and symmetrical polyelectrolytes, i.e. polycation and polyanion with identical charges and chain lengths. Especially, Nakajima et al. have reported the free energy of mixing (ΔF) for a three component system composed of water, polymer salt and microsalt;

$$\Delta F = RT[\psi_w \cdot \ln\psi_w + (\psi_p/r) \cdot \ln(\psi_p/2) + \psi_s \cdot \ln(\psi_s/2)$$
$$- \alpha(\alpha_s + \sigma \cdot \psi_p)^{3/2} + \chi_{wp} \cdot \psi_w \cdot \psi_p + (\chi_{ps}/r) \cdot \psi_p \cdot \psi_s \qquad (20)$$
$$+ \chi_{sw} \cdot \psi_s \cdot \psi_w]$$

where χ_{ij} is the free energetic interaction parameter between species i and j, ψ_i the volume fraction of species i, r the volume ratio of macroion to water, σ the charge density of the macroion, and α the electrostatic interaction parameter (see Eq. 21). The subscripts w, p and s designate water, polymer salt and microsalt, respectively.

$$\alpha = \frac{e^2}{3\,\varepsilon} \left(\frac{4\,\pi e^2}{\varepsilon kTv} \right)^{1/2} \frac{1}{kT} \tag{21}$$

ε = dielectric constant,   e = elementary charge

k = Boltzmann constant,   v = molecular volume

The theoretical values are in good agreement with experimental results using a model system, e.g. partially sulfated poly(vinyl alcohol) and partially aminoacetalyzed poly(vinyl alcohol)[102]. Polderman[109] determined the change in the free energy of mixing of oppositely charged polyelectrolytes from the difference of the results of the potentiometric titration between the theoretically calculated values obtained from individual polyelectrolyte systems assuming no interaction, and the experimental results obtained from their mixing systems by the following relation:

$$-\Delta G(pH) = R \cdot T \cdot n_{PB} \cdot \sigma_{PB} \cdot \phi_{PA} \tag{22}$$

where n_{PB}, σ_{PB} and ϕ_{PA} denote the mole of repeating units, the charge density of the polybase and the electrostatic potential of the polyacid, respectively.

3.2.1 Formation of Polyelectrolyte Complexes

The formation of polyelectrolyte complexes (PEC) is governed by the characteristics of the individual polyelectrolyte components (e.g. properties of ionic sites – strong or weak electrolyte –, position of ionic sites, charge density, rigidity of macromolecular chains) and the chemical environment (e.g. solvent, ionic strength, pH and temperature). Polyelectrolyte complexes are either separated from the solution as solids or liquids or they are still soluble in solution or may settle as gels due to variation of the controlling factors mentioned above.

The characteristics of polyelectrolyte complexes existing as combinations of strong polyacids and strong polybases were studied first. In 1961 Michaels et al.[49] reported that poly(4-vinylbenzyl-trimethylammonium chloride) (PVBMA) and poly(sodium styrenesulfonate) (NaSS) formed an equimolar complex, and since then they have extensively studied this type of complex and its applications[50-53]. As shown in Fig. 3, the conductivity of mixed aqueous solutions of PVBMA and NaSS shows a maximum at the mixing ratio, [NaSS]/([NaSS] + [PVBMA]) in mol of repeating unit/l, of 0.5. In contrast to polyelectrolytes, the molar conductance of the released microsalt (NaCl) is much higher. Therefore, the optimum mixing ratio for this polyelectrolyte complex is found to be unity. In this case, it is considered that as both polyelectrolytes are strong polyelectrolytes the degree of dissociation of each polyelectrolyte is always almost unity irrespective of the pH and their concentrations. From this result, it is concluded that immediately after at least one ionic bond is formed,

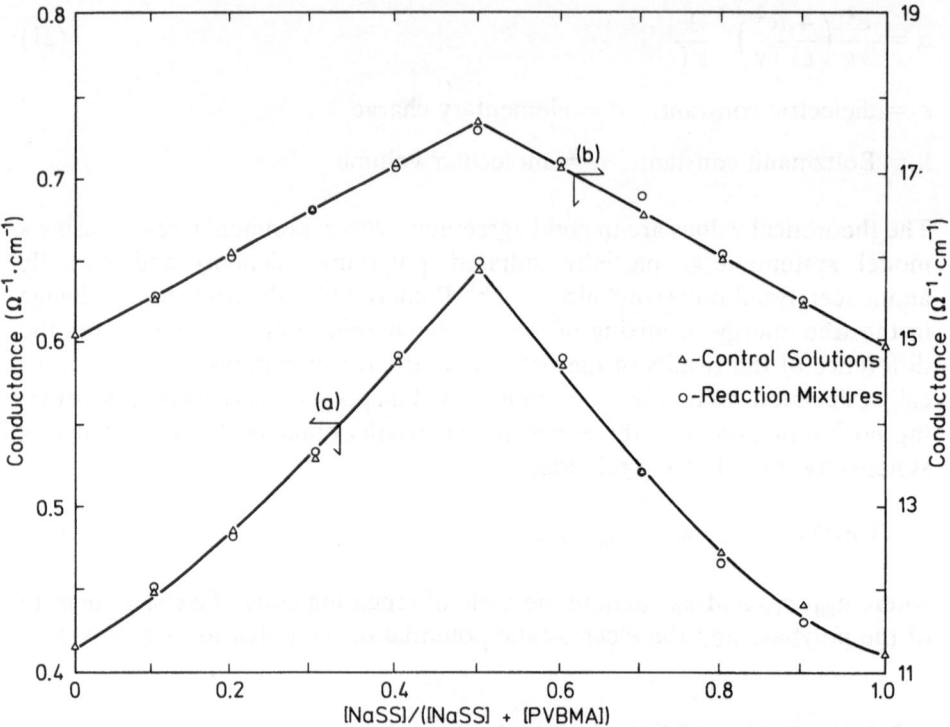

Fig. 3 a, b. Composition of polyelectrolyte complexes detected by conductometry in the poly-(sodium styrenesulfonate) (NaSS)-poly(4-vinylbenzyl-trimethylammonium chloride) (PVBMA) system[49]. (a) [NaSS] = [PVBMA] = 10^{-3} mol of repeating unit/l; (b) [NaSS] = [PVBMA] = 10^{-2} mol of repeating unit/l, [NaCl] = 10^{-2} N

adjacent reactive sites interact with complementary units at the nearest distance, thus generating an equimolar scrambled salt structure when the macromolecular chain is relatively easily coiled. This salt is insoluble in water and in common organic solvents. This phenomenon is well explained by "cooperative interaction" (for details see Sect. 3.2.2).

On the other hand, when integral-type polycations (having cationic charges on the backbone chain) instead of pendant-type polycations, are used as strong polybases, different phenomena are observed. Tsuchida et al.[86] discussed the interactions between NaSS and a series of integral-type polycations, called ionenes:

$$\left[\begin{matrix} \overset{\displaystyle CH_3}{\underset{\displaystyle CH_3}{N^+}} X^- & \overset{\displaystyle CH_3}{\underset{\displaystyle CH_3}{N^+}} X^- \\ {-N^+}{-R_1}{-N^+}{-R_2} \end{matrix}\right]_n$$

$X^-; Cl^-, Br^-$

$$\left(\begin{matrix} \text{m, m}-\text{ionene; } R_1 = R_2 = -(CH_2)_m \\ \text{m, m}'-\text{ionene; } R_1 = -(CH_2)_m\,, R_2 = -(CH_2)_{m'} \\ \text{m, X}-\text{ionene; } R_1 = -(CH_2)_m\,, R_2 = -CH_2C_6H_4CH_2- \\ \text{X, X}-\text{ionene; } R_1 = R_2 = -CH_2C_6H_4CH_2- \end{matrix}\right)$$

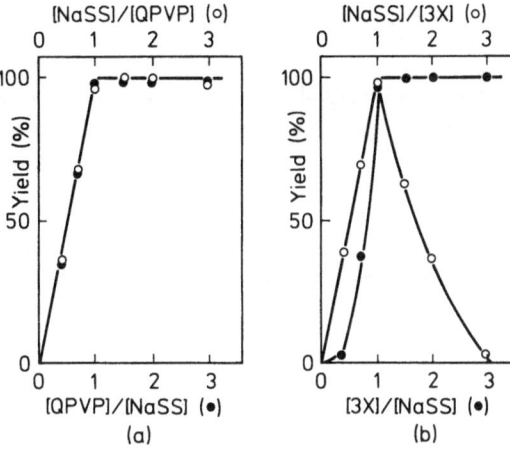

Fig. 4a, b. Effect of the structure of the polycation component on the yield of polyelectrolyte complexes. (**a**) Pendant-type polycation (QPVP)-NaSS; (**b**) Integral-type polycation (3 X)-NaSS; ○ NaSS solution added to polycation solution, ● Polycation solution added to NaSS solution

Figure 4 shows the relationships between the yields of polyelectrolyte complexes and the mixing ratios of their components. Using a pendant-type polycation (Fig. 4(a)), the yield of the complex reaches a maximum (about 100%) at a mixing ratio of 1 independent of the mixing order. In other words, the complex can always be obtained as an equimolar complex. On the other hand, when NaSS is added to an integral-type polycation (3 X) (Fig. 4(b), open circles), the yield reaches a maximum at a mixing ratio of 1 in the same manner as with a pendant-type polycation. However, further addition of NaSS causes a linear decrease of the yield, and finally the complex is redissolved completely at [NaSS]/[3 X] = 3. In the case of reverse addition order, i.e. addition of 3 X to NaSS (Fig. 4(b), closed circles), the complex is not precipitated at ratios less than [3 X]/[NaSS] = 1/3, and then its yield increases linearly until an equimolar mixture is reached. These facts demonstrate that in the region for which [3 X]/[NaSS] > 1, an equimolar water-insoluble polyelectrolyte complex is obtained whereas in the presence of excess polyanions (NaSS), a water-soluble complex with the composition [3 X]/[NaSS] = 1/3 is formed. This implies that the effect of the position of the cationic site (pendant or integral type) on the complex formation may be expressed by the complexation scheme shown in Fig. 5. Compared with the pendant-type polycation, cationic sites of the integral-type polycation are only slightly hindered. Thus, when using pendant-type polycation, and an equimolar complex is once formed, excess NaSS cannot attack the cationic sites in the complex from the opposite direction. In systems of

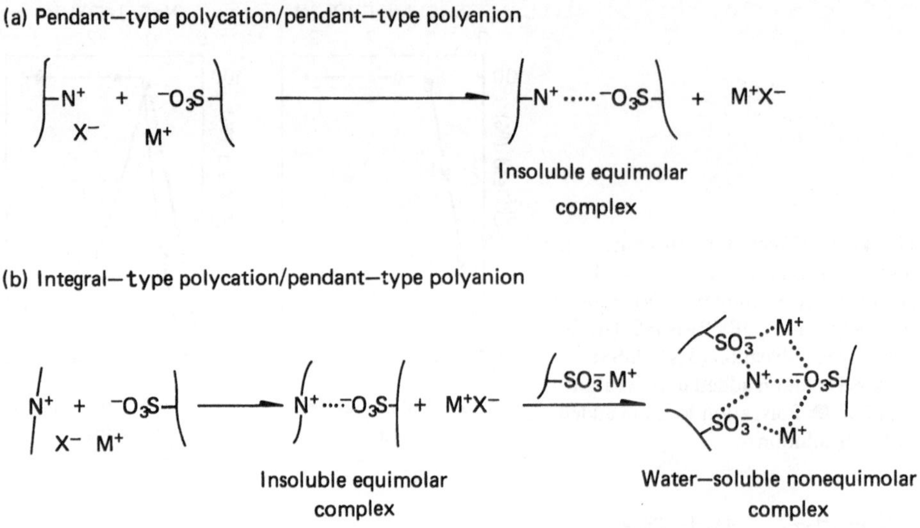

(a) Pendant—type polycation/pendant—type polyanion

(b) Integral—type polycation/pendant—type polyanion

Fig. 5. Schematic representation of the effect of the position of cationic sites of polycations on the formation of polyelectrolyte complexes

integral-type polycations, excess NaSS can attack the cationic sites of the equimolar complex from other directions to form a soluble complex with the ratio of the components of 3. In this system, counterions of NaSS, i.e. Na^+, may weaken the electrostatic interactions between polyelectrolytes. As a result, the interaction between oppositely charged polyelectrolytes also becomes less pronounced and hydration increases, and the complex becomes soluble in water.

Another interesting result obtained from these findings is that the yield is proportional to the amount of each polymer component added. Furthermore, even when the degrees of polymerization of the two polymer components differ, are identical phenomena observed. This result suggests an „all or none type" complex formation mechanism:

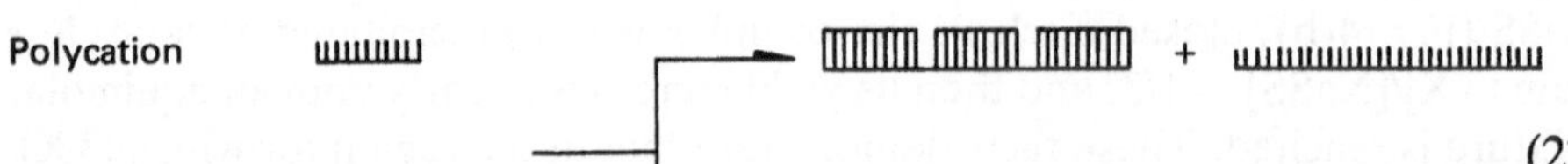

$$(23)$$

This implies that the reactivity of the polyanion chain partially covered by the polycation may be considered to be higher than that of the free chain, probably owing to the changes of conformation, dissociation and microenvironment in the doamin of the polymer chain. Therefore, completely neutralized polyelectrolyte complexes and completely free polyelectrolytes coexist in the solution (for detail see Sect. 3.2.2).

Table 4. Composition of the complexes of poly(carboxylic acid)s with various polycations in water

Poly(carboxylic acid)	Polycations	Composition*
PAA	QPVP	2 : 1
	PLL · HBr	2 : 1
	PDMAEMA · HCl	2 : 1
	5,6-ionene	3 : 1
	2,5-ionene	2 : 1
	10,10-ionene	1 : 1
	2 X	4 : 1
	3 X	4 : 1
PMAA	PVBMA	2 : 1
	10,10-ionene	9 : 1
	2 X	5 : 1
	3 X	5 : 1
PIA	2 X	3 : 1
PGA	3 X	6 : 1

Composition*: molar ratio of repeating unit ([poly(carboxylic acid)]/[polycation])
The degree of neutralization of poly(carboxylic acid)s are 0.
PAA = poly(acrylic acid), PMAA = poly(methacrylic acid), PIA = poly(itaconic acid), PGA = poly(L-glutamic acid), QPVP = poly(N-ethyl-4-vinyl-pyridinium bromide), PLL = poly(L-lysine), PDMAEMA = poly(2-N,N-dimethylaminoethyl methacrylate), PVBMA = poly(4-vinylbenzyl-trimethylammonium chloride), 5,6-ionene, 2,5-ionene, 10,10-ionene, 2 X and 3 X (see text)

Using a weak polyacid or a weak polybase as the polyelectrolyte component, the dissociation state of the weak polyelectrolyte directly affects the
complexation ability and the composition of the obtained complex. Table 4
compiles the various compositions of PEC consisting of a pair of a strong
polybase and a weak polyacid (the degree of neutralization $\bar{\alpha}$ of the weak
polyacid is zero). Poly(methacrylic acid) (PMAA) forms complexes with varying composition ([poly(carboxylic acid) unit in mol/l]/[polycation unit in mol/l]
= 1/2 ∼ 1/9) according to the structure of the polycation because of the following reasons:
(1) the charge densities of the polycations being different, the induced effects
of the polycations on the dissociation of poly(carboxylic acid) vary
(2) the hydrophobicity of polycation chains affects the stability of the complex
(3) the effect of stereocomplementarity may occur, e.g. conformity of the
interspace of two adjacent ionic sites, steric hindrance around ionic sites and
conformation and/or configuration of chains. When $\bar{\alpha}$ is zero, poly(carboxylic
acid)s scarcely dissociates in solution. However, the polycation can induce the
dissociation of poly(carboxylic acid) up to about 20–25% (assuming that nearly
all dissociated ionic sites can react with complementary active sites). Here,
especially quite interesting phenomena are observed using 10,10-ionene as the
polycation component. While the composition of PMAA-10,10-ionene is 9 : 1,
that of poly(acrylic acid) (PAA)-10,10-ionene is 1 : 1. This may be attributed to
the fact that since 10,10-ionene and PMAA, which are different from PAA,

exhibit a relatively rigid compact structure due to their high hydrophobicity, the number of real active sites becomes very low. In other systems, in general, when the charge density of the polycation increases, these sites tend to induce effectively the dissociation of poly(carboxylic acid) and to form a complex with a composition close to unity. Figure 6 shows the potentiometic titration curves of PMAA in the absence and presence of polycations, 2 X and XX, and of their low molecular weight analogs, 2 B and BT, (a) and the ratios of the proton concentration in the absence and presence of these polycations, (b). In the presence of polycations, the pH of the PMAA aqueous solution is remarkably lowered. This phenomenon is caused by the release of protons accompanied by the formation of the polyelectrolyte complexes. On the other hand, in the presence of low molecular weight analogs, the pH changes only slightly, which means that a stable complex is not formed in these systems. From these results, it can be supposed that the chain length of the polymer component is one of the most important factors for the control of the complex formation. As shown in Fig. 6(b), only in the PMAA system that differs from the PAA system, is a large peak at about $\bar{a} = 0.1$ of the neutralization of PMAA observed. It is known that the dissociation of PMAA is greatly depressed in the low pH region by its comparatively rigid conformation. Thus, the appearance of a large peak at $\bar{a} = 0.1$ means that polycations can effectively eliminate intrachain hydrophobic interactions of PMAA, resulting in facile dissociation of PMAA. In this region, XX is more effective than 2 X, because the hydrophobicity of XX is stronger than that of 2 X. However, in the region of higher degrees of neutralization, the formation of polyelectrolyte complexes is mainly affected by the electrostatic potential, i.e. the charge density, of poly-

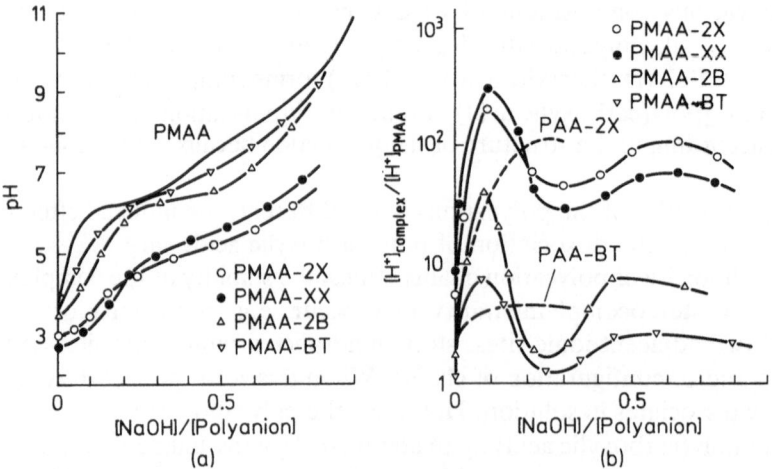

Fig. 6 a, b. Acceleration of dissociation of poly(carboxylic acid)s according to the complex formation with various polycations. (a) Potentiometric titration of poly(methacrylic acid) (PMAA) and its complexes; (b) Amount of protons released from poly(carboxylic acid) in the formation of polyelectrolyte complexes. Polyanions: PMAA and poly(acrylic acid) (PAA). Polycations and their low molecular weight analogues:

BT; $CH_3-\overset{\overset{\displaystyle CH_3}{|}}{\underset{\underset{\displaystyle CH_3}{|}}{N^+}}-CH_2-$⬡ Cl^- , 2B; ⬡$-CH_2-\overset{\overset{\displaystyle CH_3}{|}}{\underset{\underset{\displaystyle CH_3}{|}}{N^+}}-CH_2-CH_2-\overset{\overset{\displaystyle CH_3}{|}}{\underset{\underset{\displaystyle CH_3}{|}}{N^+}}-CH_2-$⬡ Cl^- Cl^-

2X; $+\overset{\overset{\displaystyle CH_3}{|}}{\underset{\underset{\displaystyle CH_3}{|}}{N^+}}-CH_2-CH_2-\overset{\overset{\displaystyle CH_3}{|}}{\underset{\underset{\displaystyle CH_3}{|}}{N^+}}-CH_2-$⬡$-CH_2\}_n$, XX; $+\overset{\overset{\displaystyle CH_3}{|}}{\underset{\underset{\displaystyle CH_3}{|}}{N^+}}-CH_2-$⬡$-CH_2-\overset{\overset{\displaystyle CH_3}{|}}{\underset{\underset{\displaystyle CH_3}{|}}{N^+}}-CH_2-$⬡$-CH_2\}_n$

cations. Therefore, in this region, 2 X forms complexes more readily than XX because the charge density of 2 X is higher. From these results, it is supposed that both hydrophobic interactions and Coulomb forces affect the formation of polyelectrolyte complexes in aqueous medium. The hydrophobicity of polycations was studied by Tsuchida et al.[406] by means of fluorescence measurements. They found that the strength of hydrophobicity of integral-type polycations increases in the order $10,10 > 8,8 > 6,6 \simeq XX \simeq 6X > 2X > 4,4$-ionenes, and that it rises because of
(1) the increase of the number of methylene groups or the presence of xylylene groups between two adjacent cationic sites
(2) the presence of benzyl groups on side chains
(3) the increase of molecular weight, especially taking into account a certain critical chain length for the formation of a hydrophobic domain in polymer chains
(4) the contracted conformation of polyelectrolyte chains. Abe established that hydrophobicity is also increased by the formation of intermacromolecular complexes[407].

Table 5. Dissociation constants (pK$_a$) of PMAA and its complexes

Sample	pK$_a$	n'
PMAA	7.3	2.3
PMAA-BT	6.3	2.2
PMAA-2B	6.4	2.2
PMAA-2X	4.3	1.4
PMAA-3X	5.5	
PMAA-XX	4.8	
PMAA-PEO ($\overline{M_w}$ = 1300)	7.5	
PMAA-PEO ($\overline{M_w}$ = 25000)	7.9	
MAA*	3.5	1.0

MAA*; Methacrylic acid monomer
Henderson-Hasselbach equation
$pH = pK_a - n' \cdot \log[(1 - \bar{a})/\bar{a}]$
PMAA = poly(methacrylic acid), PEO = poly(ethylene oxide), BT, 2B, 2X, 3X and XX see Fig. 4

The degree of induction of the dissociation of PMAA in the presence of various polycations is estimated by its apparent dissociation constant (pK_a) calculated from the potentiometric titration results as shown in Fig. 6, using the Henderson-Hasselbach equation (see Table 5). In the presence of polycations, pK_a and one of the interaction parameters relative to adjacent ionic sites, n', decreases. This confirms the induction of dissociation. Furthermore, the difference in the composition of the polyelectrolyte complex composed of one kind of polycation and different poly(carboxylic acid)s, may be affected by pK_a and the conformation of them.

In order to describe quantitatively the steric effect on the complexation, Valuyeva et al.[90] studied the interaction between a weak polybase, poly(2-N,N-dimethylaminoethyl methacrylate) (PDMAM), copolymers of a strong acid monomer (sodium styrenesulfonate: SSNa), and two kinds of nonionic comonomers; one type is 3-o-methacrylic ester of 2,4-di-o-tosyllaevoglucosan (DTMLG) with large bulkiness, and the other methyl methacrylate (MMA) with low bulkiness. The degree of conversion (Θ) decreases linearly with increasing substitution by nonionic groups (x). This means that the effect of bulkiness of the comonomer can be estimated by a constant b called "screening coefficient", usig the following equation:

$$\Theta = \Theta_0 - bx \tag{24}$$

The value of b is found to be 1.2 for DTMLG and 0.2 for MMA. This fact is explained by the difference of the bulkiness of the introduced nonionic comonomers (steric hindrance).

When the degree of neutralization (i.e. the pH of the solution) of a weak polyelectrolyte is changed, the composition of the polyelectrolyte complexes containing a polycation is varied as in the resulting systems of some poly(carboxylix acid)s-ionenes shown in Fig. 7. On the assumption that the polyelectrolyte complexes are formed upon complete neutralization, their compositions r are denoted by the following equations:

$$[PC] = [PA] \cdot \alpha \tag{25}$$

$$r = [PC]/[PA] = \alpha \tag{26}$$

where [PC] and [PA] mean the molar concentrations of polycation and polyanion (in this system poly(carboxylic acid)), respectively, and α is the degree of dissociation of PA in the presence of the PC (calculated from pK_a in Table 5), providing that the degree of dissociation of the PC is unity. Using PAA, the experimental values of r are in good agreement with the theoretical values (Fig. 7). In the case of PMAA or poly(L-glutamic acid) (PGA), however, the experimental values deviate from the theoretical values. This may be due to the specific and drastic conformational changes of both poly(carboxylic acid)s, i.e. the changes of the equilibria, packed coil ⇋ random coil for PMAA and α-helix ⇋ random coil for PGA.

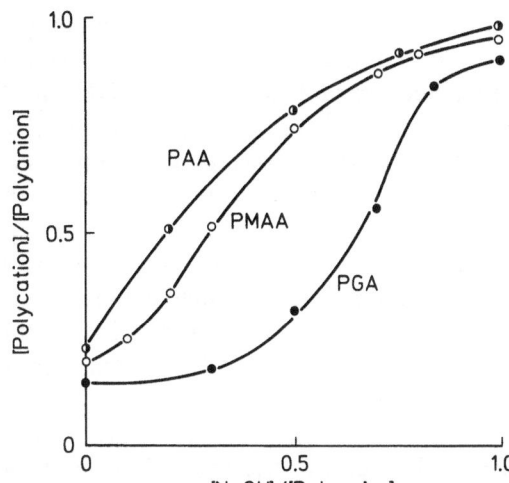

Fig. 7. Effect of the conformation of the polyelectrolyte components on the composition of the complexes. PMAA = poly(methacrylic acid), PAA = poly(acrylic acid), PGA = poly(L-glutamic acid), Polycation = 3 X (integral-type polycation)

In the system of a weak polyacid and a weak polybase, whose degrees of dissociation depend on the pH of the solution, the composition of the obtained complex is as follows:

$$[PA] \cdot \alpha_{PA} = [PC] \cdot \alpha_{PC} \tag{27}$$

$$r = [PC]/[PA] = \alpha_{PA}/\alpha_{PC} \tag{28}$$

This relationship has been examined for the systems of poly(vinylpyridine) (PVP)-PGA[508] or partially carboxymethylated and aminoacetalyzed poly-(vinyl alcohol)[152–157]. In these systems, as shown in Fig. 8, only in the neutral pH region where both weak polyelectrolyts can partially dissociate, is the complex formed, mainly as a complex coacervate[508]. However, at higher or lower pH, the coomplex is not formed because the degree of dissociation of

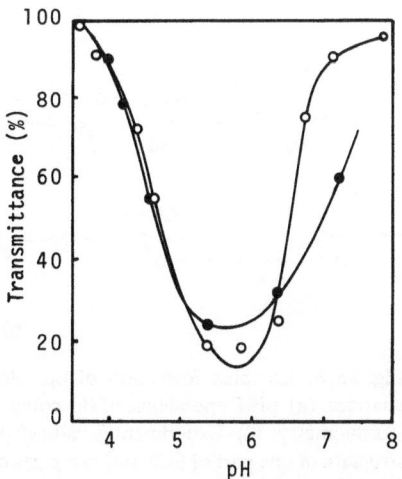

Fig. 8. pH Dependence of the complexation ability of the combination of a weak polyacid with a weak polybase.
○ = poly(L-glutamic acid)-poly(4-vinylpyridine); ● = poly(L-glutamic acid)-poly(2-vinylpyridine); Solvent; water-methanol (50 : 50 by volume)

either the weak polyacid or the weak polybase is relatively low. Even if partial formation of the complex occurs, is it soluble due to sufficient solvation of the unreacted ionizable groups, i.e. $-NH_2$ or $-COOH$. On the other hand, Naka-jima et al.[509-511] reported that if polyelectrolytes with more rigid polymer chains such as polysaccharides were used, the composition in some cases derived from stoichiometry. They found that the glycol chitosan-heparin system showed stoichiometry whereas the glycol chitosan-sulfated cellulose and -hyauronic acid systems did not, as shown in Fig. 9. A model of a ladder like complex composed of sulfated cellulose and glycol chitosan is also illustrated in Fig. 9. This result was interpreted by taking into account

(1) the conformations of the polymer components, e.g. correlative positions of complementary reactive sites

(2) the distribution of ionizable groups along the chains

(3) rigidity, e.g. restriction of internal rotations and bonds, of the chains.

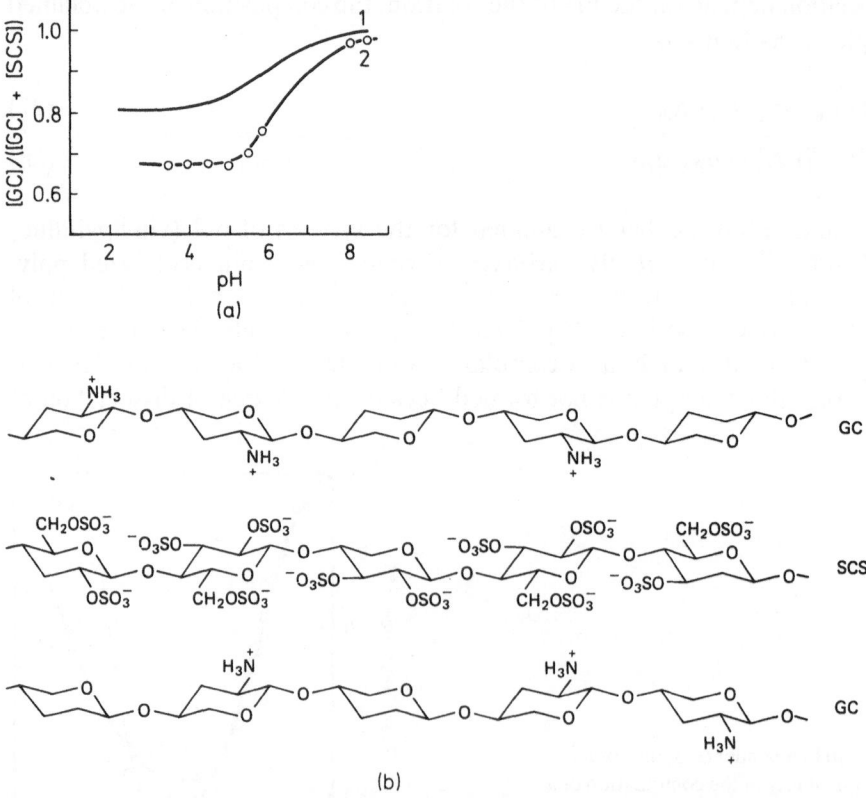

Fig. 9 a, b. Complex formation of polyelectrolytes with rigid polymer chains such as polysac-charides. (**a**) pH Dependence of the composition of the complex. (1) Theoretical values assuming stoichiometry; (2) Experimental values; (**b**) Schematic representation of a ladder-like complex structure of one part of SCS and two parts of GC; SCS = Sulfated cellulose, GC = glycol chitosan

As a consequence, the complex formation is considered to cause the conformational changes of macromolecular chains. As typical examples, the conformational changes by complexation of various polymers with specific conformations, e.g. α-helix, β-sheet structure, double-stranded helical structure, etc. have been studied. Table 6 summarizes the conformations of basic homopolypeptides forming complexes with various polyanions. Poly(L-lysine) and poly(L-arginine) may change their conformations to an α-helix, a β-form, a super-helix, a triple helix or a random coil, according to the structure of the polyanion, i.e. conformation, charge density, position of charge, and rigidity, and to the length of side chains of homopolypeptides. On the other hand, PGA, an acidic homopolypeptide, is in general rich in random coil structures due to complex formation with various polycations, although there is a difference in the degree of destabilization of α-helical structures[89, 434] caused by complementary polymers (Fig. 10). The effect of configuration, e.g. tacticity, on the complex formation was studied by Nakajima et al.[404] in poly(L-lysine) (PLL)-isotactic (iso)-, syndiotactic (synd)-, conventional- and atactic (at)-PMAA systems. As shown in Fig. 11 the helical content of PLL in the complex varies with the configuration of PMAA. Iso-PMAA with high regularity induces the formation of the α-helical structure, although conventional PMAA

Table 6. Conformation of basic homopolypeptides in polyelectrolyte complexes with various polyanions

Polyanion	Poly(L-lysine)	Poly(L-arginine)	Ref.
Poly(L-glutamic acid)	–	α-Helix	412
Poly(L-glutamic acid)	β-form	–	413, 433
	Superhelix (pH < 3)	–	414, 415
Poly(D-glutamic acid)	Random, β-form (pH 11)	–	414
Poly(L-aspartic acid)	–	Partly α-helix	412
Poly(L-aspartic acid)	Random	–	413
Poly(acrylic acid)	Partly α-helix	α-Helix	89, 412
Poly(methacrylic acid)	α-Helix	–	416
Polyphosphate	Partly α-helix	–	417
Poly(vinylsulfonate)	–	Partly α-helix	412, 418
Poly(sodium styrenesulfonate)	Partly β-form	–	419, 421
Mucopolysaccharides	α-Helix	α-Helix	422, 426
Poly(U)	–	Random	412
Poly(U)	Random	–	422
Poly(I + C)	Triple helix	–	427
Poly(A + U)	Triple helix	–	428
RNA	Random	–	422, 427
DNA	Random	Random	412
	Triple helix	–	429, 432
dDNA	Random	Random	412, 422

Poly(U) = poly(uridylic acid), poly(I + C) = equimolar complex of poly(inosinic acid) and poly(cytidylic acid), poly (A + U) = equimolar complex of poly(adenylic acid) and poly(uridylic acid), RNA = polyribonucleic acid, DNA = polydeoxyribonucleic acid, dDNA = denatured DNA

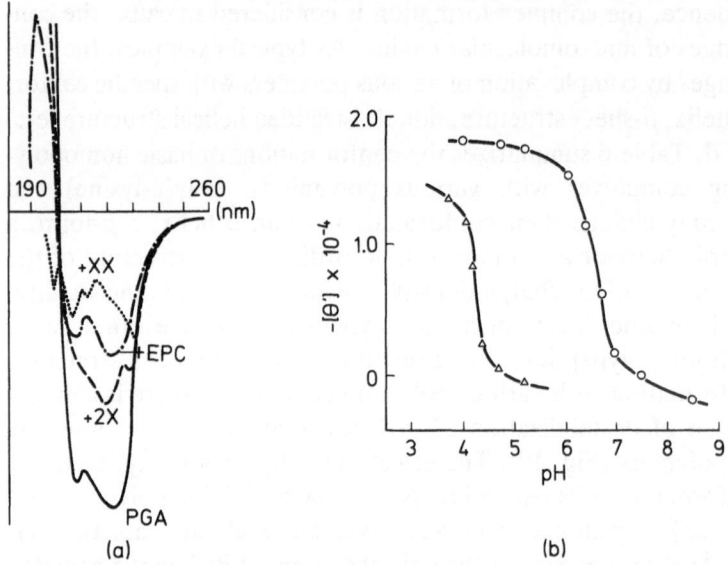

Fig. 10 a, b. Conformational change of polymer components upon complexation
(a) CD spectra of poly(L-glutamic acid) (PGA)-polycation complexes; 2 X and XX are Ionene-type polycations, EPC = poly(oxymethyl-1-methyelene-N,N,N-trimethylammonium chloride);
(b) Helix-coil transition of PGA and its complex with 6,6-ionene; ○ PGA, △ PGA-6,6-ionene complex

with lower regularity simultanously displays both helix-inducing and -destructing effects. This helix-destructing effect is proportional to the content of atactic configuration of PMAA, i.e. iso ≪ synd < conventional ≲ at.

Izumrudov et al.[145] reported non-equimolar water-soluble polyelectrolyte complexes composed of PMAA and quaternized poly(4-vinylpyridine) (poly(N-ethyl-4-vinylpyridinium bromide); QPVP). When the pH is increased after making the mixing ratio (expressed as the molar ratio of [QPVP]/[PMAA]) equal to unity in the neutralization ($\bar{a}$) of PMAA beyond $\bar{a} = 0.8$, it is confirmed that an equimolar water-insoluble complex is formed. If the initial

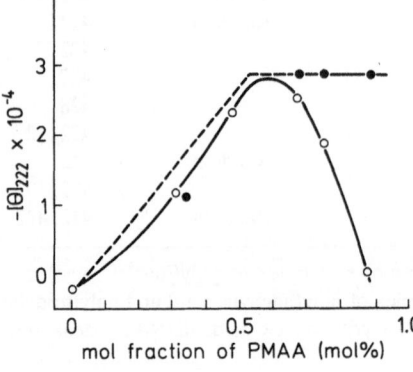

Fig. 11. Effect of tacticity on the complex formation of poly(L-lysine)(PLL)-poly(methacrylic acid) (PMAA)[414], ● PLL-isotactic PMAA, ○ PLL-conventional PMAA

mixing ratio is 1/5, a complex with this ratio is obtained at $\bar{\alpha} = 0$ similar to the above mentioned experiment (mixing ratio = 1/1). However, when increasing the pH of the solution beyond $\bar{\alpha} = 0.5$, the complex is completely redissolved. It was proposed[145] that in the range $0 < \bar{\alpha} < 0.2$, reorientation or rebinding occurs in the complex whereas at $0.2 < \bar{\alpha}$ excess carboxylic acid residues, which are not involved in the formation of electrostatic bonds, redissociate to give the surface anionic character (hydrophilic), resulting in the solubilization of the complex.

Up to now, the effects of the characteristics of the polyelectrolyte components themselves have been discussed. The effect of the reaction conditions, for example ionic strength, solvent, concentration and temperature on the formation of polyelectrolyte complexes will also be discussed in detail. When increasing the ionic strength, the following phenomena are expected to be observed:

(1) reduction of electrostatic interactions due to the screening effect of microsalts

(2) acceleration of dissociation of weak polyelectrolytes owing to the decrease of intramolecular electrostatic repulsion,

(3) increase of hydrophobicity caused by the contraction of polyelectrolyte chains.

On the other hand, the increase of concentration of the polymer components leads to the suppression of the dissociation of the polyelectrolyte components due to the rise in the electrostatic repulsion within inter- and intramacromolecules and to the interpenetration of polymer chains.

Figure 12 shows the effect of ionic strength on viscosity (a) and transmittance (b) of the mixed aqueous solution of PMAA and integral-type polycation, 2X. In contrast to the NaSS-2X complex, the PMAA ($\bar{\alpha} > 0.5$)-2X

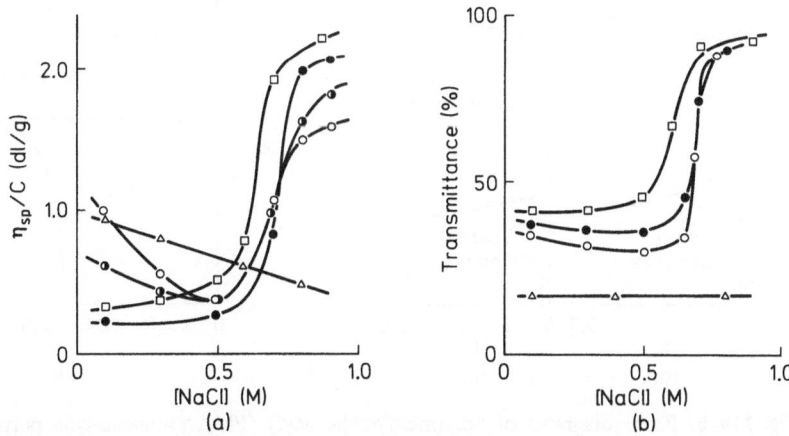

Fig. 12a, b. Effect of ionic strength on the formation of polyelectrolyte complexes (a) Reduced viscosity, (b) transmittance; ● Poly(methacrylic acid) (PMAA) (degree of neutralization of PMAA; $\bar{\alpha} = 1$)-Ionene-type polycation (2X), ○ PMAA($\bar{\alpha} = 0.75$)-2X, ◐ PMAA($\bar{\alpha} = 0.5$)-2X, □ Poly(acrylic acid) ($\bar{\alpha} = 1$)-2X, △ Poly(sodium styrenesulfonate)-2X

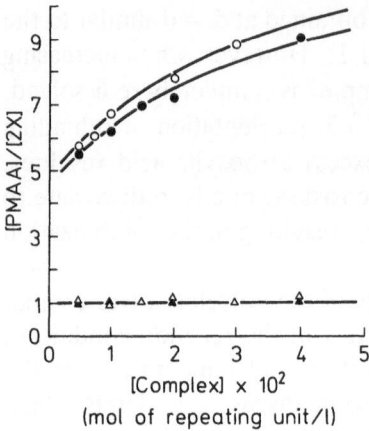

Fig. 13. Dependence of the composition of the polyelectrolyte complex of poly(methacrylic acid) (PMAA)-ionene-type polycation (2 X) on the concentration of the complex; ○ Degree of neutralization of PMAA $(\bar{a}) = 0$, ionic strength $(\mu) = 0$, ● $\bar{a} = 0, \mu = 0.01$, △ $\bar{a} = 1, \mu = 0$, ▲ $\bar{a} = 1, \mu = 0.01$

complex is dissociated into the corresponding polyelectrolyte components at 0.7 M NaCl, independent of the composition of the complex. On the other hand, at $\bar{a} < 0.5$, the complex may readily precipitate with increasing ionic strength. The effect of polymer concentration on the composition is also shown in Fig. 13. While at $\bar{a} = 1$ the composition of the complex is always unity, at $\bar{a} = 0$ the ratio of the components changes from 5 to 9, according to the increase in concentration. Figure 14 shows the phase diagrams of this complex. In dilute solution, the complex is formed, whereas in concentrated systems at low pH, it is not formed because of suppression of the dissociation of PMAA. At concentrations around 10^{-2} mol/l in repeating unit, at $\bar{a} = 0$, the complex is obtained as a curdy precipitate whereas at $\bar{a} = 1$, it separates as a complex coacervate. The effect of ionic strength in a wider pH range on complexation is shown in Fig. 14(c). A boundary line between complex coacervates and pre-

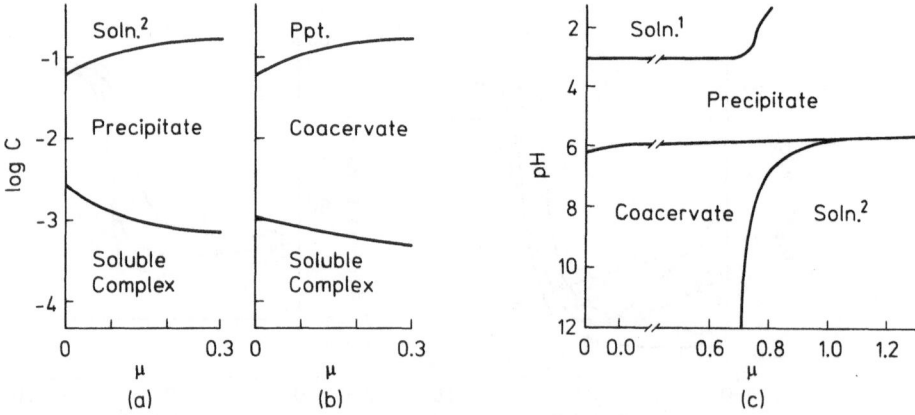

Fig. 14a, b. Phase diagrams of poly(methycrylic acid) (PMAA)-ionene-type polycation (2 X) complex. (a) and (b): Ionic strength (μ) as a function of the concentration of the complex at the degree of neutralization of PMAA $(\bar{a}) = 0$ and $\bar{a} = 1$, respectively. (c) Dependence of pH on μ at constant concentration of the complex (5 × 10^{-3} mol of repeating unit/l) at 25 °C. Solution[1] = complex is formed (soluble under this condition); Solution[2] = complex is not formed

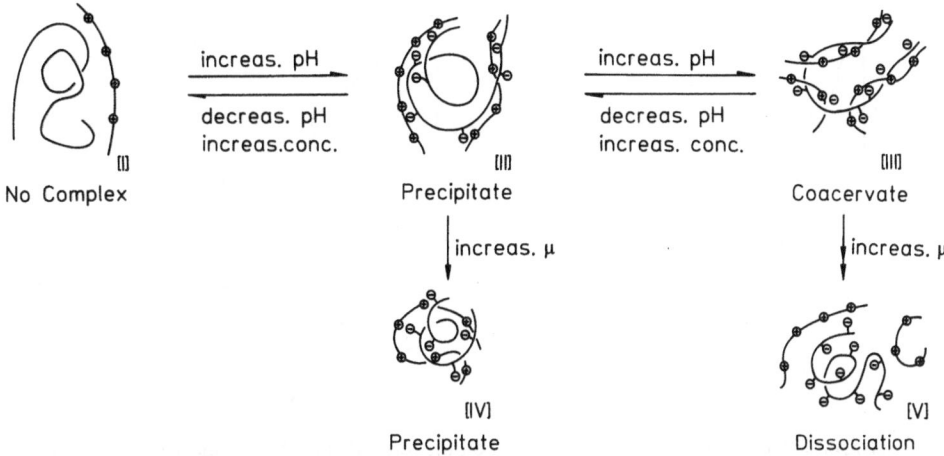

Fig. 15. Schematic representation of the equilibrium for the phase changes of the poly(methacrylic acid)-ionene-type polycation complex

cipitates is drawn at about pH = 7. At higher pH, the composition of the complex is unity. When increasing the ionic strength more than 0.7, the complex coacervates are dissociated to the individual polyelectrolyte components because of the screening effect of microsalts. In contrast to this phenomenon, the precipitate is not dissociated. Figure 15 shows the schematic diagram of such phase changes of the polyelectrolyte complexes composed of a weak polyacid and a strong polybase. PMAA cannot dissociate at extremely low pH and very high concentrations (I). Under these conditions, the complex is not formed because the amount of dissociated carboxylate anions is insufficient for the generation of a stable complex. On increasing either the pH or ionic strength, or upon dilution (II), PMAA can form a stable complex with polycations, provided the amount of carboxylate anionic sites exceeds a certain critical value. At this stage, the complex is obtained as a curdy precipitate. A further increase of ionic strength (IV) enhances the contraction of the polymer chain. Thus the complex yield increases with decreasing solubility and the complex is obtained as a more compact precipitate. At a pH of the solution > 7.0, an equimolar complex coacervate is generated, due to complete dissociation of PMAA (III). In this pH range hydrophobic interactions may be diminished by the considerable hydration. When a large amount of microsalts is added to this system beyond $\mu = 0.7$, the complex coacervate is broken (V).

Since the component polyelectrolytes are difficult to dissolve in organic solvent, only few studies on the solvent effect on the formation of polyelectrolyte complexes have been reported. In organic solvents, the following phenomena are expected to be observed:
(1) Coulomb forces are strengthened but
(2) the dissociation of the polyelectrolyte components is lowered with decreasing dielectric constant (ε) of the solvent

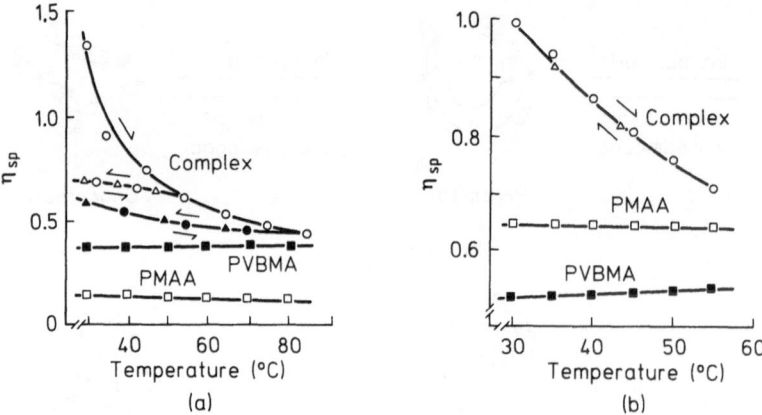

Fig. 16 a, b. Effect of temperature and solvent on the formation of the polyelectrolyte complexes of poly(methacrylic acid) (PMAA)-poly-(4-vinylbenzyltrimethylammonium chloride) (PVBMA); **(a)** In water, **(b)** in methanol

(3) hydrophobic interactions are considerably weakened. Figure 16 describes the dependence of the viscosity of PMAA-PVBMA in water and methanol on temperature. At elevated temperatures, the hydrophobic interactions are reinforced in aqueous medium whereas the Coulomb forces are only slightly changed, and the kinetic energy of the polymer chain increases. Thus, a gel is formed in both solvents, and the optimum composition ([PVBMA]/[PMAA]) is 0.5 in aqueous medium and 0.17 in methanol. The viscosity considerably decreases on heating the polyelectrolyte complex in both systems. However, it is very interesting that the viscosity change is irreversible in water and reversible in methanol. From these results, it is supposed that in water the complex is stabilized by irreversible conformational change owing to hydrophobic interactions. In contrast, the hydrophobic interactions can be neglected in organic solvents.

The intramacromolecular complexation of the block copolymer of poly(styrenesulfonic acid) (PSS)-poly(2-vinylpyridine) (P 2VP) has recently been reported. Varoqui et al.[435] have synthesized this block copolymer and investigated its solution properties. They observed a great increase in the basicity of the pyridyl groups in the block copolymer. This phenomenon was explained by invoking a stoichiometric interaction between oppositely charged blocks.

To conclude, the following phenomena can be pointed out:
(1) the composition of the complex is generally affected by the degree of dissociation of the polyelectrolyte components except the polymer components exhibit specific structures, e.g. conformations such as an α-helix and densely packed coils, as well as high rigidity, high steric hindrance around ionic sites, and position of ionic sites (i.e. integral and pendant)
(2) the charge densities of the polyelectrolyte components affect the induction of the association of ionizable units of the components (change of the composition of the complex) and also the hydration of the complex, thus controlling

Table 7. Compilation of resin preparation methods[52]

Methods	Advantages	Disadvantages	Ref.
Titration of electrolyte solution to other electrolyte solution	Stoichiometric product yield regardless of mixing ratio; no purification of starting materials required no waste of solvent	Dilute solution (<0.1%) requires excessive water; filtration may be difficult	436
Add solutions of polyelectrolytes simultaneously to drowing bath	Stoichiometric product yield regardless of mixing ratio; no purification of starting materials required	Control of product addition rates difficult; reactor mixing and product titration also troublesome	437
Coprecipitate complex from ternary solution	Working with concentrated polyelectrolytes (10% vs. 1%) possible pure compact product	Product not necessarily stoichiometric; solvent handling and disposal can be difficult, expensive solvents	438
Mechanical blend on rubber mill; ross mixer, etc.	Elimination of large volumes of water and/or undesirable solvents	Expensive equipment and skilled operators required; product not necessarily stoichiometric; filtration may be difficult	439

the phase of the complex (formation of precipitate, coacervate or soluble state)

(3) hydrophobic interactions stabilize the complex, in addition to electrostatic interactions,

(4) at relatively low ionic strength ($\mu < 0.7$), the complex may easily be formed, probably owing to the induction of the dissociation of polyelectrolytes whereas at relatively high ionic strength ($\mu > 0.7$), the coacervate complex tends to dissociate to the polyelectrolyte components because of the screening effect of the microsalts (complex precipitates are not dissociated)

(5) in organic solvents, the complex is also formed.

Resins of polyelectrolyte complexes may be prepared commercially by at least four methods compiled in Table 7[52].

3.2.2 Physical and Chemical Properties of Polyelectrolyte Complexes

To achieve a wider commercial application of polyelectrolyte complexes, more detailed data on their physical and chemical properties are required. It is generally assumed that polyelectrolyte complexes have unique properties because the main interaction forces are the strong Coulomb forces and their electrostatic nature (net charge) can easily be varied by changing only their composition[51]:

(1) physicochemical properties (insolubility in common solvents, infusibility, plasticizability by water and electrolytes, highly specific but limited water-absorption)

(2) good transparency as a wetted flexible solid
(3) selectivity of ion-sorption and ion-exchange properties
(4) electrical properties (high dielectric constant and loss factor in the wetted state sensitive to moisture and ion content, low d.c. conductance up to rather high electrolyte-dopant levels, extremely high d.c. conductance in contact with concentrated aqueous electrolytes)
(5) transport properties (high permeability to water, electrolytes and other water-soluble microsolutes, inpermeability to macrosolutes),
(6) good anti-coagulant properties.

Polyelectrolyte complexes composed of a strong polyacid and a strong polybase are insoluble in common organic and inorganic solvents. They are only soluble in specific ternary solvent mixtures, i.e. water/water-compatible organic solvents/microsalts (e.g. water/acetone/NaBr). The solubilities of polyelectrolyte complexes of NaSS-pendant and integral-type polycations are shown in Fig. 17. The obtained results suggest that organic solvents may weaken the hydrophobic interactions and microsalts the electrostatic interactions between macromolecules (see Fig. 14). Other useful ternary solvents are mixtures such as water/acetone/sulfuric acid, water/ethanol/hydrophobic acid and water/dioxane/calcium dichloride.

Michaels et al.[50-53] reported the chemical and physical properties of the polyelectrolyte complex of NaSS-PVBMA. This complex is used as a membrane in various fields. Polyelectrolyte complexes prepared by casting homogeneous solutions of ternary-solvent systems mentioned above are transparent and amorphous. Dried polyelectrolyte complexes are hard materials whereas wetted ones are rubber-like or skin-like. The polyelectrolyte complex with an equimolar composition is neutral whereas that with a non-equimolar composition can display ion-exchange properties. The range of the water content depends on the excess of polycation or polyanion in the complex.

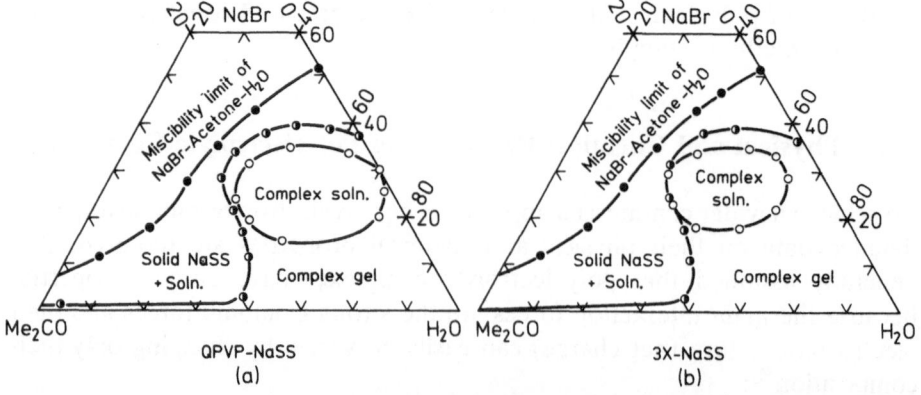

Fig. 17a, b. Solubility of polyelectrolyte complexes in ternary solvent systems at 30 °C.
(a) Poly(N-ethyl-4-vinylpyridinium bromide) (QPVP)-poly(sodium styrenesulfonate) (NaSS),
(b) Poly(N,N,N',N'-tetramethyl-p-xylylenepropyrenediammonium dichloride) (3 X)-NaSS

Polyelectrolyte complexes have extremely high and controllable permeability to water and low molecular weight solutes. For example, if the water content is the same, the water permeability is more than 10 times higher than that of reconstituted cellophane and also considerably higher than that of crosslinked poly(2-hydroxyethyl methacrylate) hydrogels. Tables 8 and 9 compile the permeabilities of water and low molecular weight solutes and Fig. 18 shows the permeability of water as a function of the gel-water content and that of the solutes as a function of their molecular weight. The permeability of low molecular weight solutes through complex membranes is higher than through the commercial cellophane membrane. It should be especially noted that complex membranes exhibit to a certain extent permeability to solutes with comparatively high molecular weight (higher than 1000). Smolen et al.[70] reported the similarity between biological lipid membranes and polyelectrolyte complex membranes with regard to the permeability of bio-related materials as shown in Fig. 19. They explained this phenomenon by the assumption of a specific structure of water in the membrane. The magnitude of oxygen permeability is quite high compared with almost any other synthetic polymer, with the exception of silicon rubber[65]. Moreover, it is of great interest that while the ratio of CO_2 to O_2 permeability is between 3 and 6 for almost all polymers that of

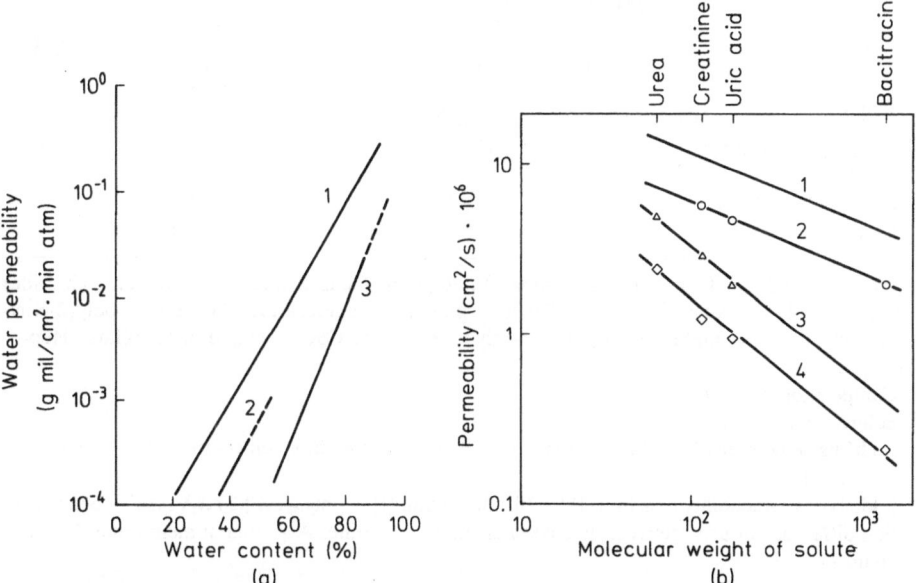

Fig. 18 a, b. Typical permeabilities of various hydrogels to water and various solutes; (a) Water permeability at pressures less than 7×10^7 dyne/cm^2 [53] (1) = polyelectrolyte complex of poly-(sodium styrenesulfonate) (NaSS)-poly(4-vinylbenzyltrimethylammonium chloride) (PVBMA), (2) = crosslinked hydrogel of poly(2-hydroxyethyl methacrylate), (3) = cellulose; (b) Dialytic permeability of a polyelectrolyte complex composed of NaSS-PVBMA to solutes with various molecular weights[54] (1) Water, (2) neutral polyelectrolyte complex (water content = 70%), (3) anionic polyelectrolyte complex (water content = 61%), (4) cellophane and cuprophane (water content = 57%)

Table 8. Water permeability of polyelectrolyte complexes

Sample	Thickness (μm)	H$_2$O (%)	K $\times$ 10^{16} (cm^2)	P	Ref.
Ioplex		66.6	0.818		56
PVBMA-NaSS	9.0	67	55.22		55
Biolon (Neutral)		57		125 $\times$ 10^8 [a]	65
Biolon (Anionic)		57		14 $\times$ 10^8 [a]	65
Ioplex (Neutral)		1.3 (g/g)		50 $\times$ 10^5 [b]	52
Ioplex (Anionic)		1.3 (g/g)		8.5 $\times$ 10^5 [b]	52
PEPP-PAA	64		0.0233		115

[a] cm^3/s $\cdot$ dyne, [b] cm^3/s at 6.9 $\times$ 10^6 (dyne/cm^2)
Ioplex and Biolon are commercial names of polyelectrolyte complexes composed of poly(4-vinyl-benzyltrimethylammonium chloride) (PVBMA) and poly(sodium styrenesulfonate) (NaSS). PEPP and PAA are poly(ethylenepiperazine) and poly(acrylic acid), respectively

Table 9. Permeability of polyelectrolyte complexes to solutes

Membrane	NaCl	Urea	Glucose	Sucrose	Ref.
Ioplex (Neutral)	0[b] (11.0)[c]	0 (12.0)	–	10 (–)	
Ioplex (Anionic)	0 (11.0)[c]	0 (29.0)	–	5 (–)	52,53
Cellophane	0 (16.8)	0 (16.8)	–	15 (–)	
PAA-PEPP (0.5:0.5)[a]	–	[10.8][d]	–	–	115
PAA-PMAEM (0.5:0.5)	–	[10.5]	–	–	
GC-CSC (0.65:0.35)	–	3.2	54.5	73.0	440
GC-Hep (0.62:0.38)	–	5.5	55.0	65.5	

Ioplex = poly(sodium styrenesulfonate)-poly(4-vinylbenzyltrimethylammonium chloride) (0.5:0.5), PAA = poly(acrylic acid), PEPP = polyethylenepiperazine, PMAEM = poly(2-N,N-dimethylaminoethyl methacrylate), GC = glycol chitosan, CSC = chondroitin sulfate, Hep = heparin,
[a] Composition of the complex
[b] Solute retention (%)
[c] ( )diffusive permeability, P ($\times$ 10^7) in cm^2/s (P = Dk$_2$, D = diffusion coefficient, k$_2$ = partition coefficient)
[d] [ ]dialytic permeability constant, P ($\times$ 10^3) in cm/min (P = [ln(C$_0$/ΔC)/2 At]V, ΔC = concentration difference of urea between two compartments, t = time, A = area of membrane, V = cell volume)

polyelectrolyte complexes is about 20 (approximately the same as the selectivity for CO$_2$ over O$_2$ in water). This fact indicates that the complex gel acts on permanent gases like an "immobilized water" membrane similar to the alveoli of the lungs. High permeability is a manifestation of the ionic network structures of a polyelectrolyte complex. That is, compared with non-ionogenic or ion-exchange membranes, polyelectrolyte complex membranes having both

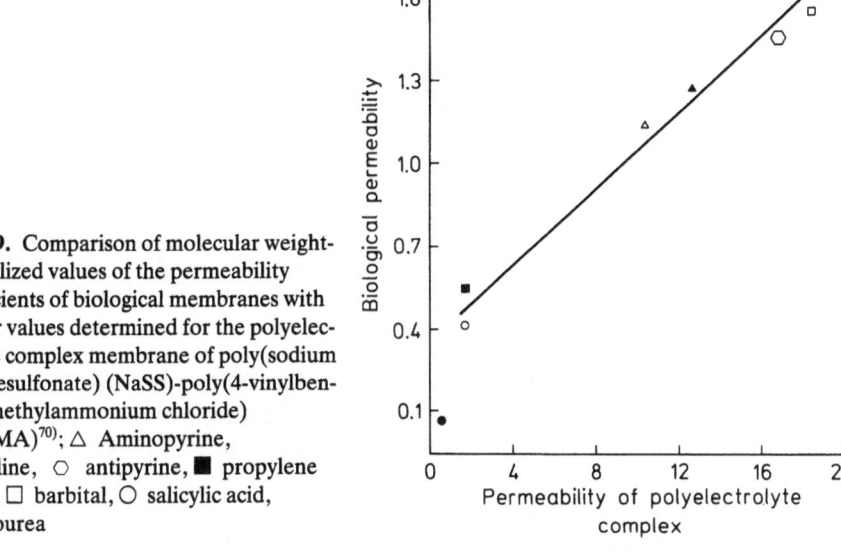

Fig. 19. Comparison of molecular weight-normalized values of the permeability coefficients of biological membranes with similar values determined for the polyelectrolyte complex membrane of poly(sodium styrenesulfonate) (NaSS)-poly(4-vinylbenzyltrimethylammonium chloride) (PVBMA)[70]; △ Aminopyrine, ▲ aniline, ○ antipyrine, ■ propylene glycol, □ barbital, ○ salicylic acid, ● thiourea

anionic and cationic sites, can effectively destroy the structure of water in the membrane, and selective permeability is caused by the above mentioned ability to control the gel-water content and ionic crosslinking topology.

Membranes made from polyelectrolyte complexes based on the combination of a weak polyacid and a weak polybase, e.g. PAA-poly(ethylenepiperazine) (PEPP) with an equimolar composition, exhibit a relatively high permeability to water and urea[115]. The dialysis constants of the permeability to urea (P) are listed in Table 10, and Fig. 20 shows the permeability coefficients for water (K) of the membranes as a function of the imposed pressure (ΔP). Figure 20 shows that though the coefficient of permeability of cellulose remains constant at any pressure, investigations of the ultrafiltration properties of polyelectrolyte complex membranes can only be made within a limited range of applied pressures. This means that this membrane is characterized by pressures ΔP_{Hm} at which the coefficients of permeability increase

Table 10. Dependence of the dialysis constant for the permeability of urea (P) on the composition of the polyelectrolyte complex membrane[115]

	[PAA]/[PEPP]				Cellulose
	1	3	5	10	
Thickness (nm)	60	35	85	90	43
s^a (%)	50	30	40	73	40
P $\times$ 10^3 (cm/min)	10.8		9.8	15.5	13.7

[a] Water content: PAA = poly(acrylic acid), PEPP = polyethylenepiperazine

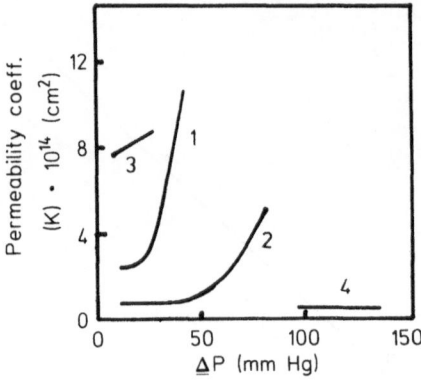

Fig. 20. Dependence of the permeability coefficient of the polyelectrolyte complex of poly(acrylic acid) (PAA)-polyethylenepiperazine (PEPP) for water on the imposed pressure[115]. (1) [PAA]/[PEPP] = 1/1, (2) [PAA]/[PEPP] = 3/1, (3) [PAA]/[PEPP] = 10/1, (4) Cellulose

sharply. Stretching of the membrane is thus observed, resulting in a further rise in ΔP. The increase in the permeability coefficent may be explained by the growing interstructural spacing in the polyelectrolyte complex as a result of planar orientation during stretching, which is equivalent to an increase in the effective pore size in the membranes. The observed stretching of the membrane does not disappear after removal of the imposed pressure. It may also be seen from Fig. 20 that a change in the polyelectrolyte complex composition leads both to a variation in ΔP_{Hm} and to a change in the absolute values of the permeability coefficients. As stated above, polyelectrolyte complex membranes have specific permeabilities different from other commercial membranes, and one can easily vary the permeability only by selection of the polyelectrolyte components and the compositions.

In the following, the mechanical properties of polyelectrolyte complexes will be discussed. In Table 11, the tensile strength, modulus, and elongation of the polyelectrolyte complexes of NaSS-PVBMA at medium water content and PAA-PEPP are compared with the corresponding parameters of cellophane and a slightly crosslinked poly(2-hydroxyethyl methycrylate) hydrogel having a comparable water content. NaSS-PVBMA is stiffer than the other hydrogels

Table 11. Mechanical properties of some hydrogels at room temperature

Hydrogels	Strength[a] (dyne/cm²) × 10⁻⁷	Elongation (%)	Modulus[b] (dyne/cm²) × 10⁻⁷
NaSS-PVBMA (1:1) (55% H₂O)	5.5	18	55
PAA-PEPP (1:1) (50% H₂O)	5.0	400	3.0
PAA-PEPP (1:1) (35% H₂O)	16	250	130
Cellophane (45% H₂O)	28	92	30
Crosslinked HEMA (45% H₂O)	0.41	140	<6.9

[a] Tensile strength,   [b] Young's modulus
NaSS = poly(sodium styrenesulfonate), PVBMA = poly(4-vinylbenzyl-trimethylammonium chloride), PAA = poly(acrylic acid), PEPP = poly(ethylenepiperazine), HEMA = poly(2-hydroxyethyl methacrylate)

and intermediate in strength. The dynamic-mechanical properties of this complex have also been examined. The master stress-relaxation curve for a polyelectrolyte complex is similar to those of conventional glassy amorphous polymers. The presence of water and of electrolytes superimposes upon the effect of temperature in the promotion of rubbery behavior. The relationships between some mechanical properties and the water content are depicted in Fig. 21. Tensile strength and modulus decrease whereas elongation increases

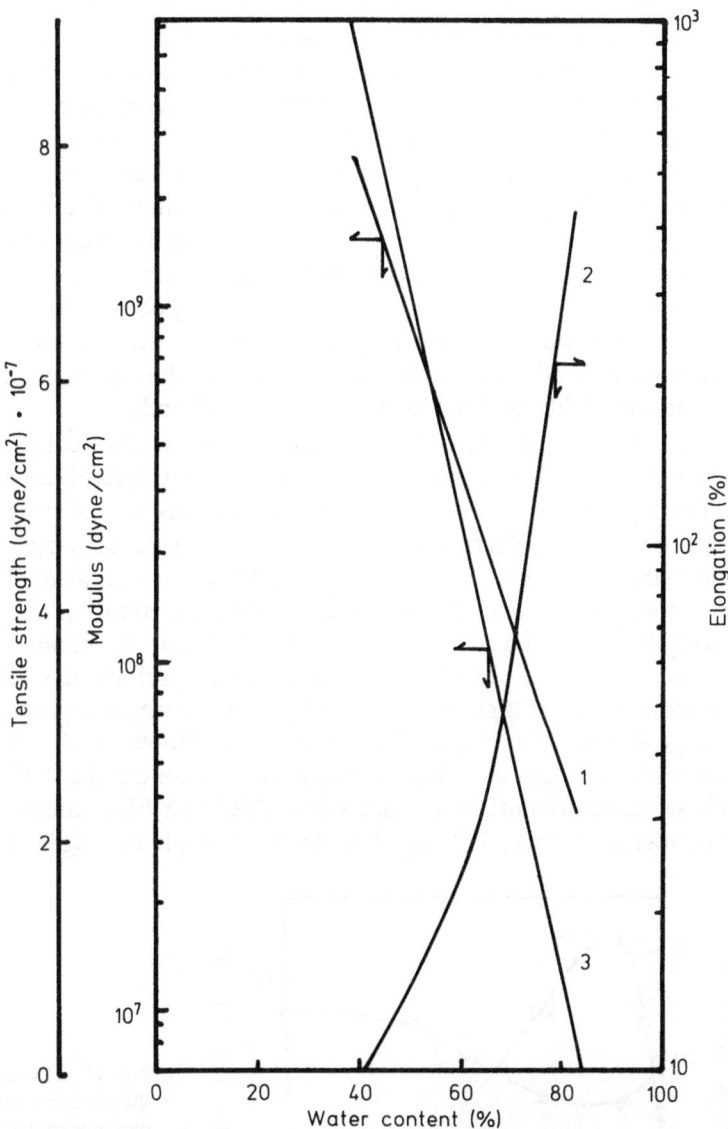

Fig. 21. Mechanical properties of the polyelectrolyte complex of poly(sodium styrenesulfonate) (NaSS)-poly(4-vinylbenzyl-trimethylammonium chloride) (PVBMA)[54]. (1) modulus, (2) elongation, (3) tensile strength

in an inverse ratio with rising water content. This fact suggests that water may act as a plasticizer for the complex. The tensile strength is lower than with other plastics but similar to that of dimethylsilicone.

PAA-PEPP membranes could not be obtained in the presence of an excess of the polycation (PEPP). The membranes with excess PEPP swell stronlgy in water and disperse spontaneously. An excess of polyanion (PAA) leads to an increase in the composition up to a ratio of [PAA]/[PEPP] = 20. It has been found that the physical properties of the complex strongly depend on its composition. Comparing the initial moduli (E_0) and the water content (S) in the membranes with the compositions investigated at a particular temperature (see Fig. 22), it is found that the curves describing the dependence of the composition on E_0 and S pass through extreme values, E_0 passing through a maximum and S through a minimum at the same membrane composition. The membranes characterized by high values of the initial modulus at 20 °C have corresponding [PAA]/[PEPP] ratios of 1.5. The values of the initial moduli fall sharply over a narrow temperature interval; on the other hand, in the case of membranes of equimolar composition, the initial moduli changes linearly with temperature. The structure fomation in the polyelectrolyte complex membranes based on weak pairs of polyelectrolytes, when an excess of the polyanion is introduced, may be due to the nature of the excess polymer component, i.e. in the ability of PAA to form hydrogen bonds.

In order to elucidate the difference in the mechanical properties of polyelectrolyte complexes using weak or strong polyelectrolytes as polymer components, Nakajima et al.[440, 441] reported the mechanical properties using as a pair of weak polyelectrolytes various poly(vinyl alcohol) derivatives (carboxymethylated (PA) and aminoacetalyzed (PC) poly(vinyl alcohol)) , and as a pair of strong polyelectrolytes sulfated poly(vinyl alcohol) (PSA) and poly(vinyl alcohol) acetalized with 2,2-diethoxyethyl-trimethylammonium (PTC), as shown in Table 12. The PTC-PSA complex is insoluble in water but soluble at extremely low or high pH and at high ionic strengths, probably because the charge density of each polyelectrolyte is low. However, the PC-PA complex is insoluble in aqueous solution which can solubilize the PTC-PSA complex. These results demonstrate that while PTC and PSA interact mainly due to electrostatic forces, PC and PA do so through hydrogen bonds as well as

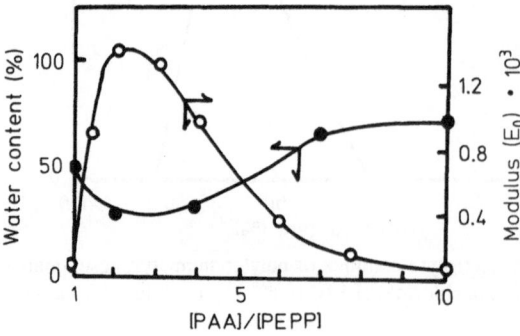

Fig. 22. Effect of composition on the water content (S) and the initial modulus (E_0) of the polyelectrolyte complex of poly(acrylic acid) (PAA)-poly(ethylenepiperazine) (PEPP)[115]

Table 12. Mechanical properties of the complexes of aminoacetalized poly(vinyl alcohol) (PC)-carboxymethylated poly(vinyl alcohol) (PA) and sulfonated poly(vinyl alcohol) (PSA)-poly(vinyl alcohol) acetalized with 2,2-diethoxyethyltrimethylammonium (PTC)[441]

Sample	Condition of heat treatment		In air, 65% R.H., 20°C			In water, 20°C		
	Temp. (°C)	Time (min)	Tenacity (dyne/cm²) $\times 10^{-8}$	Elongation (%)	Modulus (dyne/cm²) $\times 10^{-9}$	Tenacity (dyne/cm²) $\times 10^{-6}$	Elongation (%)	Modulus (dyne/cm²) $\times 10^{-5}$
PVA[a]	No	–	3.5	220	2.1	–	–	–
	200	10	4.6	140	5.0	–	–	–
PC-PA[b]	No	–	2.1	201	0.83	1.4	179	2.8
	140	60	4.3	117	5.7	27.6	24.6	1170
PTC-PSA[b]	No	–	3.2	139	5.6	2.8	600	0.7
	200	10	4.0	192	8.1	5.5	250	30

[a] Poly(vinyl alcohol) fraction with a degree of polymerization of 1120

[b] Stoichiometric complexes

through electrostatic forces. Compared with poly(vinyl alcohol) (PVA), the tensile strength of PTC-PSA is higher whereas that of PC-PA is lower. Their tensile strength in water is lower than that in air.

The electrical properties of polyelectrolyte complexes are more closely related to those of biologically produced solids. The extremely high relative dielectric constants at low frequencies and the dispersion properties of salt-containing polyelectrolyte complexes have not been reported for other synthetic polymers. Neutral polyelectrolyte complexes immersed in dilute salt solution undergo marked changes in alternating current capacitance and resistance upon small variations in the electrolyte concentration. In addition, their frequency-dependence is governed by the nature of the microions. As shown in

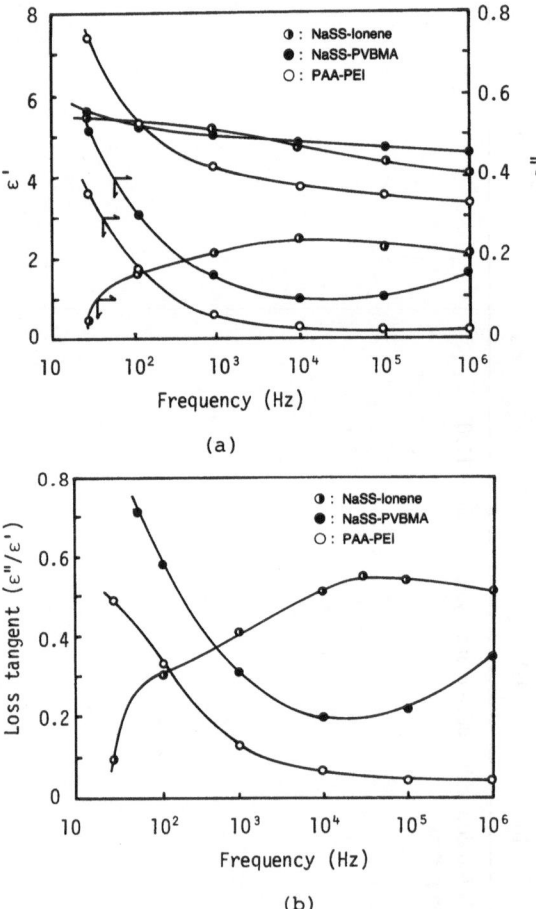

(a)

(b)

Fig. 23 a, b. Dielectric properties of polyelectrolyte complexes[88]; (a) Variation of dielectric constant ε' and loss factor ε'' of polyelectrolyte complexes with frequency; (b) Variation of loss tangent $\varepsilon''/\varepsilon'$ of polyelectrolyte complexes with frequency. NaSS = poly(sodium styrenesulfonate), PAA = poly(acrylic acid), PVBMA = poly(4-vinylbenzyl-trimethylammonium chloride), PEI = polyethyleneimine, Ionene (Bubond 60) = $[-O(CH_2)_2N^+(CH_3)_2(CH_2)_2N^+(CH_3)_2(CH_2)_2 \cdot 2\,Cl^-]_n$

Fig. 23, the polyelectrolyte complexes exhibit relatively low dielectric constants (ε') and loss factors (ε'') which slowly decrease with increasing frequency except for the PAA-polyethyleneimine (PEI) system. Hence, the loss tangent ($\varepsilon''/\varepsilon'$) monotonously decreases with frequency. Moreover, even if polyelectrolyte complexes contain a certain amount of microsalt, is the direct current conductance low. This dielectric behavior has been ascribed to the polarizability of the electrolyte sorbed into isolated microscopic domains within the matrix of the polyelectrolyte complexes.

The refractive index for a series of homogeneous polyelectrolyte complexes in the range of 40–80% gel water content at 22 °C is given by

$$n = 1.294 + 0.336 (1 - \alpha) \tag{29}$$

where α is the gel-water content expressed as a weight fraction[52].

3.3 Complexes Formed by Hydrogen Bonding

A hydrogen bond is formed through intermolecular interaction between an electron-deficient hydrogen and a region of high electron density. Its fundamental role in the structure of DNA and the secondary and tertiary structures of proteins is known. The specific properties and structures of water are also caused by hydrogen bonds.

Many intermacromolecular complexes governed through hydrogen bonds occur in biological systems as mentioned previously. However, complexes between synthetic polymers are limited to only a few systems consisting of proton-donating and proton-accepting polymers, e.g. poly(carboxylic acid)s-poly(ethylene oxide) (PEO), -poly(vinyl alcohol) (PVA) and -poly-(N-vinyl-2-pyrrolidone) (PVPo). As nucleic acid models, many studies have also been made on the interactions between synthetic polymers containing nucleic acid bases (uracil (U), cytosine (C), thymine (T), adenine (A), inosine (I), and guanine (G)), e.g. between a pair of polyribonucleic acid (poly(A)-poly(U)[442–453], poly(I)-poly(C)[454–463] and vinyl-type polymers[464–470].

3.3.1 Formation of Hydrogen-Bond Containing Complexes

Proton-accepting polymers and proton-donating polymers typically interact with each other in aqueous medium and organic solvents almost stoichiometrically. This complex formation is affected by temperature, polymer structure, polymer concentration, solvent and other interaction forces, e.g. hydrophobic interactions. In general, the ratio [proton-accepting polymer units]/[proton-donating polymer units] in mol/l of the complex is almost unity in dilute solu-

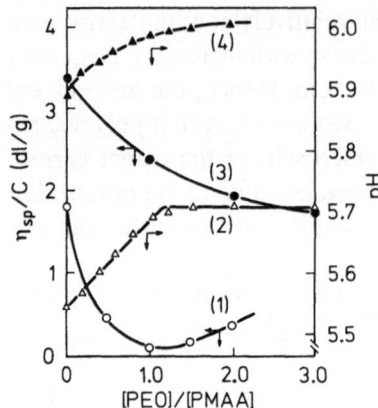

Fig. 24. Effect of pH on the formation of intermacromolecular complex through hydrogen bonds in poly-(methacrylic acid) (PMAA)-poly(ethylene oxide) (PEO) system. (1) and (2): pH of the initial mixing solution = 5.5. (3) and (4): pH of the initial mixing solution = 5.9

tion[1] and 2/3 in concentrated solution. Using poly(carboxylic acid) as a proton-donating polymer, the complexation strongly depends on the pH of the aqueous solution; i.e. on the degree of dissociation of poly(carboxylic acid). Dissociation of poly(carboxylic acid) is suppressed in the presence of PEO and is estimated by the apparent dissociation constant (pK_a) (see Table 5). From specific viscosity and pH measurements of the mixed solution of PMAA and PEO (see Fig. 24), it is found that there is a critical state of dissociation called "critical pH" (which corresponds to a "critical chain length" as stated in Chap. 4). The complexation mechanism involving hydrogen bonds is described by Eq. (30) and (31):

$$
\begin{array}{l}
\text{-COOH} \\
\text{-COO}^{\ominus} \\
\text{-COOH} \\
\text{-COOH}
\end{array}
+
\begin{array}{l}
\text{O} \\
\text{O} \\
\text{O} \\
\text{O}
\end{array}
\xrightleftharpoons[\text{below critical pH}]{}
\begin{array}{l}
\text{-COOH--O} \\
\text{-COOH--O} \\
\text{-COOH--O} \\
\text{-COOH--O}
\end{array}
\tag{30}
$$

$$
\begin{array}{l}
\text{-COO}^{\ominus} \\
\text{-COOH} \\
\text{-COO}^{\ominus} \\
\text{-COO}^{\ominus}
\end{array}
+
\begin{array}{l}
\text{O} \\
\text{O} \\
\text{O} \\
\text{O}
\end{array}
\xrightleftharpoons[\text{above critical pH}]{}
\begin{array}{l}
\text{-COO}^{\ominus} \quad \text{O} \\
\text{-COOH--O} \\
\text{-COO}^{\ominus} \quad \text{O} \\
\text{-COOH---O}
\end{array}
\tag{31}
$$

The existence of a certain number of undissociated carboxy groups is necessary for PMAA and PEO to form a stable complex through hydrogen bonds. If this condition is satisfied at a certain pH (so-called "critical pH"), a stable complex is formed irreversibly. In such complex formation, dissociated carboxy groups are influenced by the complexation and become undissociated by the extrac-

---

1 Tsuchida et al.[222] reported that in aqueous solution, the ratio of the components of the system of PMAA-PVPo is 1 : 1 whereas Bekturov et al.[223] found a ratio of 3 : 2. This discrepancy may be caused by the influence of the concentration, pH and molecular weight of individual polymer components

tion of protons from the solution into the domain of the polymer chains. Thus, most carboxy groups of PMAA are able to participate in complexation, the composition of the complex [PEO]/[PMAA] being equal to unity. At higher pH values where such conditions cannot be satisfied, as the number of active sites is extremely insufficient, it is assumed that the enthalpy afforded by hydrogen bonds is not compensated by the decrease in entropy. Hence, the equilibrium reaction is dominant over complex formation. Cooperativity in complexation and a critical pH value are also revealed by the fact that the yields of the precipitated complex change drastically at a definite pH. This critical pH depends mainly on the facility of dissociation of poly(carboxylic acid)s. For example, it is about 5.7 for PMAA, about 5.2 for an alternating copolymer of styrene and maleic acid (PSMA) and about 4.8 for PAA; pK_a (PMAA) = 7.3, pK_a (PSMA) = 6.5 and pK_a (PAA) = 5.6.

The solvent effect is considered to be one of the most important controlling factors because a certain solvent can also interact with polymers via hydrogen bonds. Therefore, this complex formation can be regarded as a three-component reaction involving a proton-donating polymer, a proton-accepting polymer and a solvent. Figure 25 shows the relationship between the ability of formation of the complex of PMAA and PVPo and the dielectric constant (ε) of the solvent (a). The parameter $C_{\tau = 50}$, which denotes the polymer complex formation ability, indicates the polymer concentration at which the polymer complex solution shows 50% transmittance[222]. In aprotic solvent, the polymer complex formation is impaired by the increase of ε. It is well-known that the static dielectric constant is affected by temperature. In general, on lowering the temperature, ε increases. In the case of DMF, ε is 52.1 at $-40\,^\circ$C. At this temperature, the polymer complex is not formed in DMF. On the basis of

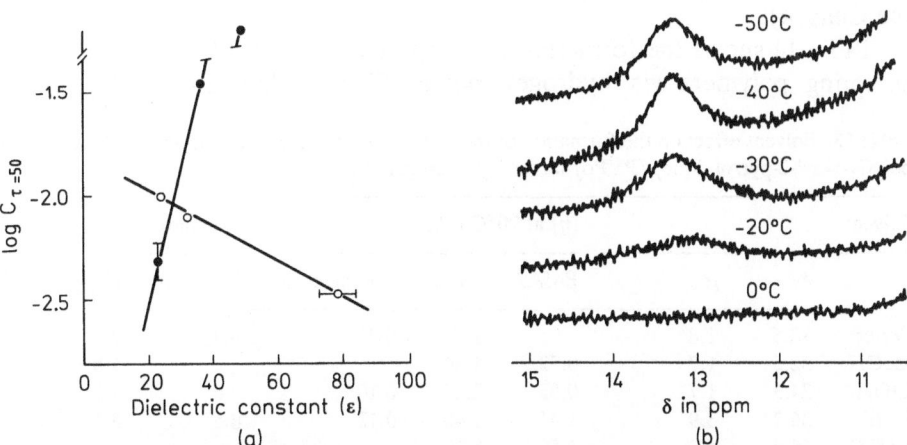

Fig. 25 a, b. Solvent effect on the formation of the poly(methacrylic acid) (PMAA)-poly(N-vinyl-2-pyrrolidone) (PVPo) complex formed by hydrogen bonds. (**a**) Relationship between the complexation ability and dielectric constant ε of solvents. ○ Protic solvents, ● aprotic solvents; (**b**) Temperature dependence of the broadening of the ^{1}H-NMR carboxy proton peak of pMAA in dimethylformamide (DMF)

these results, the dielectric constant of the solvent may be used as a parameter of the strength of the interaction between the polymer components and solvent. The interaction between PMAA and DMF is detected by means of NMR spectra (Fig. 25 (b)). The signal of the carboxy proton of PMAA, which indicates strong hydrogen-bonding interactions between PMAA and DMF, is not detected at room temperature. It is, however, observed at ca. 13 ppm at relatively low temperature and at 12.5 ppm in dimethylsulfoxide (DMSO) at room temperature. These results indicate that when the interaction between PMAA and solvent is stronger than that between PMAA and PVPo, a polymer complex is not formed. In the case of protic solvents, the opposite tendency is observed as indicated by open circles in Fig. 25 (a). This behaviour may be caused by the ability of the formation of hydrogen bonds between solvent molecules. Thus, when the dielectric constant of the solvent is increased, the hydrogen-bond forming ability between solvent molecules becomes stronger, resulting in a decrease of the interaction forces between solute and solvent.

Rough values of the molecular weights of the complexes formed in different solvents are compiled in Table 13. The complex generated in aqueous medium has a weight average molecular weight of about 4.4×10^6 whereas for those formed in organic solvents $\overline{M_w} \simeq 8.0 \times 10^6$. The intrinsic viscosities of such complexes formed in an aqueous medium and in organic media are about 0.4 and 0.1 (dl/g), respectively. Thus, the complexes obtained in organic solvents have higher molecular weights and lower intrinsic viscosities. From these results, it is suggested that, in aqueous medium, unreacted ionizable sites of PMAA in the complex are slightly disociated and the polymer complex, therefore, reaches a looser domain. But there still remains the question of the size of such polymer complexes. For example, the size of the polymer complex depends on the aging time, the initial polymer concentration etc. (for details see Chap. 4).

Table 14 shows the formation of complexes between PMAA and proton-accepting polymers via hydrogen bonds. Tetramethylurea (TMU), N,N-

Table 13. Solvent effect on the formation of the complex of poly(methacrylic acid) (PMAA) with poly(N-vinyl-2-pyrrolidone) (PVPo) linked by hydrogen bonds

Solvent			$[\eta]$ at 20 °C (dl/g)			Complex	
	ε	μ	PMAA	PVPo	Complex	$M_w{}^a \times 10^{-6}$	$C_{\tau = 50}{}^b \times 10^2$
Water	78.5	1.8		1.75	0.42	4.4	3.2
MeOH	32.6	1.7	0.72	1.86			7.8
EtOH	24.3	1.7	0.95	2.24	0.10	7.0	1.0
DMF	36.7	3.9	1.41	1.40	0.12	8.5	3.7
DMSO	48.9	4.0	1.50	1.10	–	–	–

[a] Molecular weight was determined by the laser light scattering method
[b] Concentration of polymer component (mol of repeating unit/l) at which the transmittance of the polymer complex solution is 50%
$[\eta]$ = Intrinsic viscosity, ε = dielectric constant, μ = dipole moment

Table 14. Solvent effect on the formation of the complex between poly(methacrylic acid) and various proton-accepting polymers

Solvent	Proton-accepting polymers			
	PFEI	PAEI	PVPo	PHMPA
TMU			Complex	
DMF	None	None	Complex	Complex
DMA		None	None	Complex
NMPo	–	–	None	Complex
DMSO	None	None	None	Complex
HMPA			None	None

TMU = tetramethylurea, DMF = dimethylformamide, DMA = dimethylacetamide, NMPo = N-methyl-2-pyrrolidone, DMSO = dimethyl sulfoxide, HMPA = hexamethylphosphoric triamide

PFEI	PAEI	PVPo	PHMPA

$$+CH_2-N-CH_2+_n$$
$$\underset{H}{\overset{\mid}{C}}=O$$

$$+CH_2-N-CH_2+_n$$
$$\underset{CH_3}{\overset{\mid}{C}}=O$$

$$+CH_2-CH+_n$$

dimethylacetamide (DMA), N-methyl-2-pyrrolidone (NMPo) and hexa-methyl-phosphoric triamide (HMPA) as well as DMF and DMSO are used as aprotic solvents having various dielectric constants. In addition, proton-accepting polymers exhibiting the same structure as some aprotic solvents (structures; see Table 14) are used. A rectangle in this table means that the structural unit of a proton-accepting polymer component is the same as that of the aprotic solvent employed. No complexes is formed for the couples designated by these rectangles. These results indicate that – even if the proton-

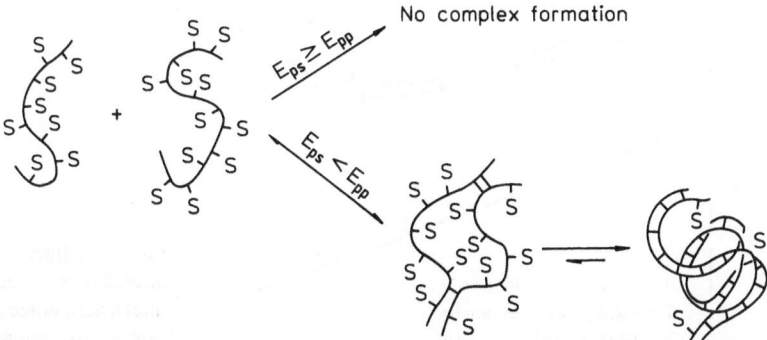

Fig. 26. Schematic representation of competitive reactions among proton-accepting polymer, proton-donating polymer and proton-accepting solvent molecules following the formation of the intermacromolecular complex through hydrogen bonds. S = Solvent molecules; E_{ps} = Interaction forces between proton-donating polymer and solvent molecules. E_{pp} = Interaction forces between proton-donating polymer and proton-accepting polymer

accepting abilities of solvent and polymer are nearly equal – solvent can interact more easily with PMAA than with the proton-accepting polymer because the amount of solvent is much larger than the proton-accepting groups of the polymers, and the mobility of the solvents is higher. The schematic diagrams of the formation of the hydrogen-bonding complex are shown in Fig. 26. The formation of this complex is assumed to have an intermediate position in the interaction between a polymer and the complementary polymer (E_{pp}) and between the polymer and solvent (E_{ps}). It should be noted that the interaction strength of a given monomeric unit in the polymer component is affected by the configurational environment of this unit and that proton-accepting forces of functional groups alone would not fully account for complex formation reactions. When E_{pp} is larger than E_{ps}, the intermacromolecular complex is formed. However, when E_{ps} is nearly equal to E_{pp} or larger than E_{pp}, no intermacromolecular complex is formed. In organic solvents, the formation of complexes through hydrogen bonds may be understood only due to such solvent effect. However, in aqueous medium, hydrophobic interactions and the dissociation state of PMAA must also be taken into consideration.

Figure 27 shows the temperature dependence of the transmittance of the solution of complexes composed of PVPo-PMAA and PVPo-PAA in water and DMF. The transmittance is measured immediately after mixing the solution of PMAA and PVPo at constant temperature, because when the complex molecules are once formed they tend to aggregate with each other in the process of time. Such aggregation phenomenon of intercomplexes is discussed in detail in Sect. 4.3. At elevated temperatures, it is expected that the kinetic energy of a polymer chain and hydrophobic interactions increase (up to 60 °C in aqueous medium) and the dielectric constant of the solvent decreases. In aqueous medium (indicated by open circles in Fig. 27), the transmittance

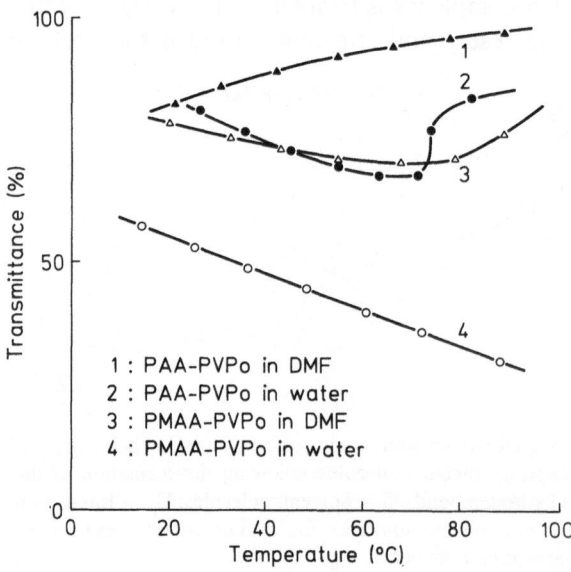

1 : PAA–PVPo in DMF
2 : PAA–PVPo in water
3 : PMAA–PVPo in DMF
4 : PMAA–PVPo in water

Fig. 27. Effect of hydrophobic interactions on the formation of intermacromolecular complex through hydrogen bonds in poly-(methacrylic acid) (PMAA)-poly(N-vinyl-2-pyrrolidone) (PVPo) and poly(acrylic acid) (PAA)-PVPo systems in water and dimethylformamide (DMF)

decreases monotonously with rising temperature in the PMAA-PVPo system. In contrast, in DMF the transmittance is recovered at higher temperature as indicated by closed circles. This fact means that the aggregate complexes become unstable at higher temperatures in DMF because the hydrophobic interactions between the polymer components are markedly weakened in organic solvent and the kinetic energy of the polymer components increases with rising temperature. In water, hydrophobic interactions become stronger with increasing temperature. In order to elucidate the effect of the α-methyl groups of PMAA on the hydrophobic interactions, the same experiment is carried out by using PAA instead of PMAA. In this system, a drastic increase in the transmittance is observed at 60 °C in water. As a result, it is revealed that hydrophobic interactions play an important role in the formation of a stable complex in aqueous medium.

In order to confirm the effect of hydrophobic interactions on the formation of the intermacromolecular complexes, calorimetric measurements on the hydrogen-bonding complex systems have been made. The process of the formation of the complex between polymer A (PA) and polymer B (PB) is described by the following scheme:

$$PA{-}Solvent \xrightarrow{\ \Delta H_1^A\ } PA + Solvent$$

$$PB{-}Solvent \xrightarrow{\ \Delta H_1^B\ } PB + Solvent \qquad \xrightarrow{\ \Delta H_2\ } PA{-}PB + Solvent \tag{32}$$

$$\xrightarrow{\ \Delta H_3\ } Complex{-}Solvent$$

$$\Delta H_1 = \Delta H_1^A + \Delta H_1^B \tag{33}$$

The first step in Eq. (32) involves desolvation. In this step, ΔH_1 may be positive (endothermic reaction) and its absolute value is related to the strength of the interaction between solvent and each polymer component. The following step is considered to involve complex formation between desolvated polymer components. In this step, hydrogen bonds are formed between polymers so that ΔH_2 may be negative (exothermic reaction). The final step consists in the conformational change involving complex formation and other factors (ΔH_3). The total heat of mixing as shown in Table 15 is the sum of the increments, ΔH_1, ΔH_2 and ΔH_3:

$$\Delta H^M = \Delta H_1 + \Delta H_2 + \Delta H_3 \tag{34}$$

In aqueous solution of the system PMAA-PVPo, ΔH^M is positive and its absolute value is 1.4 kcal/mol. Assuming that the heat of conformational change in the complex formation is the same as that of PMAA in aqueous solution (obtained as heat of dilution), namely -0.6 kcal/mol, we obtain

Table 15. Heat of mixing of proton-donating polymers and proton-accepting polymers in water and in organic solvents

System	Solvent	Conc.[c] × 10³	ΔH × 10⁻² (cal/mol)
PMAA-PVPo[a]	H₂O (pH 3)	0.2	14
PMAA-PVPo[a]	DMF	10	−2.5
PMAA-PEO[b]	H₂O (pH 3)	0.5	3.0
PMAA-PEO[b]	H₂O/MeOH = 1	10	−1.7
PAA-PEO[b]	H₂O (pH 3)	1.0	1.3
PAA-PEO[b]	H₂O/MeOH = 1	10	−1.8

[a] Temperature = 20.5 ± 0.05 °C
[b] Temperature = 25.0 ± 0.05 °C
[c] mol of repeating unit/l
PMAA = poly(methacrylic acid), PVPo = poly (N-vinyl-2-pyrrolidone), PEO = poly(ethylene oxide), PAA = poly(acrylic acid)

$$\Delta H_1 + \Delta H_2 = 1.4 - (-0.6) = 2.0 \text{ kcal/mol} \tag{35}$$

In comparison with the general hydrogen bond energy (about 5 kcal/mol), this experimental value is small. This may be attributed to the fact that not all active sites are involved in the formation of the complex and moreover, this value is calculated from Eq. (35). Considering the relation,

$$\Delta F^M = \Delta H^M - T \cdot \Delta S^M \tag{36}$$

ΔF^M must be negative so that a stable intermacromolecular complex is formed spontaneously. In this system ΔH^M is positive; hence, ΔS^M must be positive. From this result it may be concluded that hydrophobic interactions contribute to the complex formation in aqueous medium.

On the other hand, in DMF, ΔH^M is negative and its absolute value is 0.25 kcal/mol (exothermic reaction). Generally, in organic solvents such as DMF, hydrophobic interactions are negligible. Since the heats of dilution of PMAA and PVPo are nearly equal to zero in DMF, the heats of conformational change are also small. Thus, the reaction of PMAA with PVPo in DMF is governed by the enthalpy change. However, the absolute value of ΔH^M is small. From this result it follows that ΔH_1 is large, i.e. the interaction between solvent and component polymer is strong. This result is supported by the fact that the complexation ability of PMAA-PVPo is dependent on the dielectric constant of the solvent, i.e. the interaction between PMAA and DMF is relatively strong. Kabanov et al.[175] observed a similar phenomena in the case of PMAA-PEO systems but the absolute values of ΔH^M are smaller than those obtained for the PMAA-PVPo system. This difference of ΔH^M may be caused by the different hydrogen-bonding (i.e. proton-accepting) ability and hydrophobicity of PEO and PVPo.

Osada studied the enthalpy and entropy changes in the complex formation between PMAA and PEO by potentiometric titration[471]. The fraction of the binding groups of the complex (degree of conversion, Θ), the stability constant (K), and thermodynamic parameters (ΔF^0, ΔH^0 and ΔS^0) are related with each other by the following equations:

$$\Theta = 1 - ([H^+]/[H^+]_0)^2 \tag{37}$$

$$K = \Theta/C_0 \cdot (1 - \Theta)^2 \tag{38}$$

$$\Delta F^0 = - RT \cdot \ln K \tag{39}$$

$$d(\ln K)/d(1/T) = -\Delta H^0/R \tag{40}$$

$$\Delta S^0 = -(\Delta F^0 - \Delta H^0)/T \tag{41}$$

where C_0 is the initial concentration of PMAA (mol/l in repeating unit), and $[H^+]$ and $[H^+]_0$ are the proton concentrations of the PMAA solution in the presence and in the absence of the complementary proton-accepting polymers, respectively. The calculated results of various complexation systems are denoted in Fig. 28. The fact that both ΔH^0 and ΔS^0 have positive values in aqueous medium suggests the contribution of hydrophobic interactions. Extremely large values of ΔS^0 at 20 and 30 °C in the aqueous medium of the system PMAA-PEO, especially with molecular weights of 2000 and 3000 of PEO, are related to the entropy change for complexation, probably resulting in a release of solvated water molecules.

The effect of conformation of the polymer components on complexation has been discussed by using poly(amino acid)s, e.g. in the systems of poly(L-

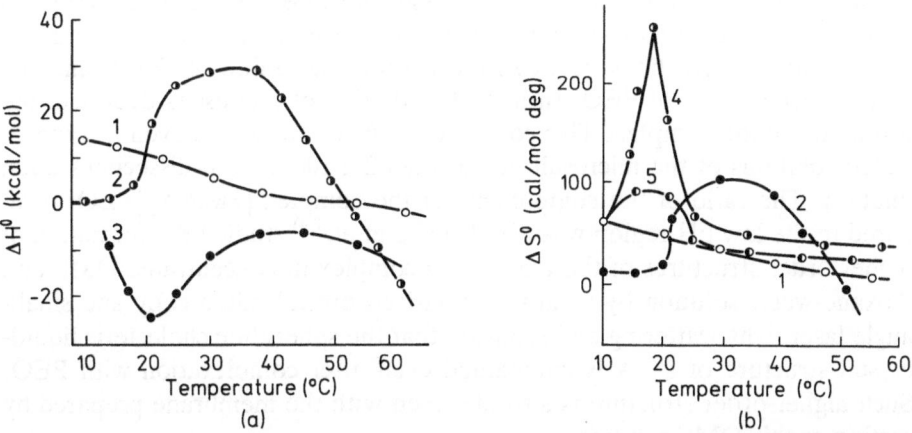

Fig. 28 a, b. Change of enthalpy and entropy in the complexation of poly(methacrylic acid) (PMAA) with poly(ethylene oxide) (PEO) of different chain length through hydrogen bonds as a function of temperature[201]. (a) Standard enthalpy change, (b) standard entropy change. (1) $M_{PEO} = 20000$, (2) $M_{PEO} = 2000$, (3) $M_{PEO} = 2000$ in ethanol-water mixture (EtOH 23%), (4) $M_{PEO} = 3000$, (5) $M_{PEO} = 7500$

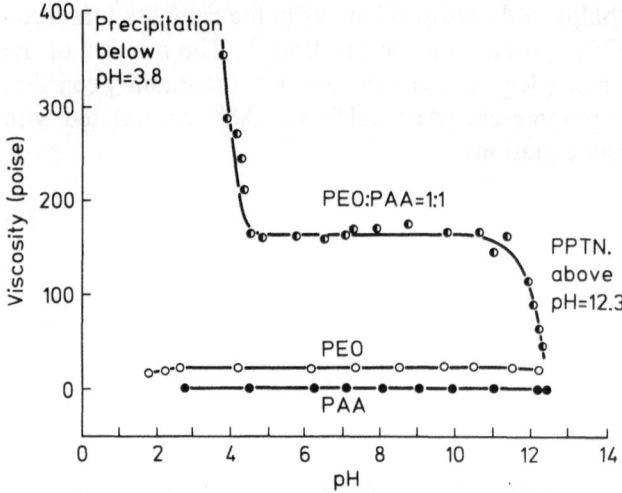

Fig. 29. The formation of intermacromolecular complex through hydrogen bonds in concentrated solutions of poly(acrylic acid) (PAA)-poly(ethylene oxide) (PEO)[176]

glutamic acid) (PGA)-PVA[472], PGA-PEO[473–478]), and poly(ε-N-carbobenzoxy-L-lysine)-PEO[479]. In general, the α-helix structure of PGA is destabilized by complexation with these proton-accepting polymers. Moreover, Kawai et al.[472] reported the growth of the β-structure of PGA in the PGA-PVA complex by aging at pH 5.7.

Bailey et al.[176] studied the complexation of PEO and PAA in concentrated system. In contrast to dilute solution systems, the viscosity of the mixed solution increases as shown in Fig. 29. This may be attributed to the formation of a network structure because compact polymer chains tend to interpenetrate. In the low pH region, as in the case of dilute solution systems, they can interact with each other through hydrogen bonds. However, it is interesting that even in the pH region where PAA partially dissociates, i.e. 4 < pH < 12, they can also form intermacromolecular complexes. Bailey et al. ascribed this behavior to ion-dipole interactions between the carboxylate anions of PAA and the ether oxygen atoms of PEO. In fact, the addition of microsalts depresses the formation of the complex. This complex is efficiently formed even in concentrated solutions of the microsalt higher than 2 g/100 ml and dissociates upon dilution. The ratio of the components of the complex [PMAA]/[PEO], prepared in the low pH region was 2/3. Tanaka et al.[474] studied the formation of higher-order structures of the PGA-PEO complex in concentrated DMF and dioxane-water solution by means of polarized optical microscopy and small-angle laser light scattering and reported that the spherulitic cholesteric liquid-crystal structure of PGA is maintained even after complexation with PEO. Such higher-order structure is also observed with the membrane prepared by casting of this DMF solution.

In heterogeneous systems, the complexation phenomena are quite different from those in homogeneous systems. The interactions between the PMAA membrane and proton-accepting polymers were studied as a function of the contraction of the membrane[177]. Figure 30 shows the profiles of the isothermal

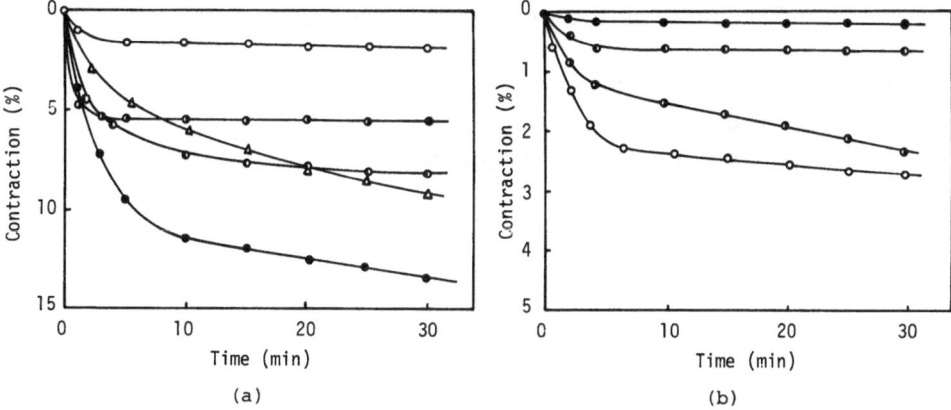

Fig. 30 a, b. Formation of intermacromolecular complexes through hydrogen bonds in heterogeneous systems[177]. (a) Time dependence of the poly(methacrylic acid) (PMAA) membrane contraction by complexation with poly(ethylene oxide) (PEO) with different chain length. O M_{PEO} = 600, ◑ M_{PEO} = 1000, ● M_{PEO} = 2000, ◐ M_{PEO} = 7500, △ M_{PEO} = 83000. (b) Time dependence of the PMAA membrane contraction by complexation with various proton-accepting polymers. O Poly(N-vinyl-2-pyrrolidone) (PVPo) (M_w = 10000); ◑ PVPo (M_w = 360000), ● poly(methyl vinyl ether) (M_w = 64000); ◐ poly(vinyl alcohol) (M_w = 22000)

contractions of the PMAA membrane observed on addition of the complementary polymers. It is obvious that PEO with a molecular weight of 2000 produces the most rapid and pronounced contraction which tends to cease after a certain period of time. PEOs with molecular weights > 2000 (e.g. M = 7500 and 83000) also show a considerable contraction over a long period of time, although contraction takes place rather slowly. In other words, PEO with high molecular weight can contract the PMAA membrane slowly but appreciably whereas PEO with low molecular weight can initially contract rapidly to some extent but equilibrium is easily attained in this process. From these results it is obvious that PEO with M = 2000 most impressively demonstrates the effect of the molecular weight of PEO on the contraction of the PMAA membrane. Another interesting feature is the fact that PEO is the only polymer which can cause a pronounced contraction of the PMAA membrane. As shown in Fig. 30 (b), PVPo, PVA, and poly(methyl vinyl ether) (PMVE) do not cause an appreciable contraction, despite the fact that all these polymers can induce marked changes of the shape of dissolved PMAA in water due to complexation. It is apparent that the contraction of the PMAA membrane cannot be simply related to the stability constant of the complexes nor to the reduction of viscosity in solution. It can, however, be related to the other factors including the chemical structure of polymers. From the result of the dependence of the molecular weight of PEO on the membrane contraction, it has been suggested that the profile of the contraction might be related to the rate of penetration of PEO into the water-swollen PMAA network. Thus, when contracting a PMAA membrane, the dimension (size) of the proton-accepting polymers in solution and the chemical structure, for example, the bulkiness and geometry of side groups, which may dominate the rate of penetration of the polymer into the membrane, should be taken into account. This assumption seems to be confirmed experimentally by the insignificant contraction obtained with PVPo, PVA, and PMVE.

It is quite interesting to study complexation systems formed by hydrogen bonds between synthetic polynucleotides as nucleic acid models. Base-base interactions are relatively easily detected by ultraviolet spectra (i.e. hypochromic effect due to the base stacking), circular dichroism, optically rotatory dispersion, infrared and nuclear magnetic resonance spectra, hydrodynamic properties (e.g. viscosity, sedimentation, etc.) and thermal properties (e.g. melting temperature, heat capacity, etc.). Studies on the interactions of homopolynucleotide, where poly(adenylic acid) (poly(A)) poly(uridylic acid) (poly(U)) form the complexes in the molar ratio of $1:1$[442] and $1:2$[443] of their monomeric units, [poly(A)]/[poly(U)], in the systems of poly(cytidylic acid) (poly(C))-poly(inosinic acid) (poly(I)) and poly(C)-poly(guanylic acid) (poly(G)), it is revealed that only one kind of complex is formed stoichiometrically[454].

The chemistry of vinyl-type polymers containing nucleic acid bases as nucleic acid models has recently been given considerable attention. A number of such synthetic polymers has been designed and synthesized, and their functions have been extensively studied because, in contrast to polynucleotides, their backbone structure exhibits the following characteristics:
(a) absence of stereoregularity,
(b) shorter distance between the various nucleic acid bases and
(c) absence of negatively charged groups[464].

Table 16 summarizes the complexes formed between the vinyl-type analogs of nucleic acid and polynucleotides[465, 466]. Poly(VA)-Poly(U) was assumed to exhibit a planar network structure. In addition, some studies on the interactions occurring in nucleic acid systems and model compounds bearing phosphate ester units on the side chains or the main chains[480], and on interactions of poly(A) and quaternized poly(4-vinylpyridine) with bases have been reported[481].

These nucleic acid model polymers strongly interact with RNA, DNA and polynucleotides but rather weakly with each other. The composition of the complex including at least one nucleic acid model polymers as a polymer component is irregular, which is in striking contrast to the stoichiometric composition of the complexes composed of polynucleotides and nucleic acids. The fact that the interactions between synthetic model polymers are weak as stated above may be attributed to structural factors (e.g. steric hindrance, conformation, distance between bases) and the strong tendency of interaction between intrachain bases. This assumption is confirmed by using linear PEI grafted L-carboxyethyladenine and L-carboxyethylthymine[467]. Moreover, Akashi et al.[468] discussed the effect of the stereoregularity of the polymer components on complexation using stereoregular methacryloyl-type polymers containing nucleic acid bases. They reported the following results: since the atactic poly-

Table 16. Complexation of synthetic polymers containing nucleic acid bases[465]

	Solvent	Poly(U)	Poly(A)	Poly(C)	Poly(G)	Poly(I)	Poly(iA)	Poly(VA)	Poly(VC)	Poly(VU)	Poly-(VHX)
Poly(VA)	H$_2$O	Complex Pu ≃ Py[e]	0	0		Complex	–[a]	O(H$^+$)			
Poly(VU)	H$_2$O	0	Complex Py ≫ Pu	0	0	0	Complex Py > Pu	Aggregate Py ≃ Pu			
Poly(VC)	H$_2$O/PG[b]	0	0	0	Complex Py > Pu	Complex Py > Pu			O(H$^+$)		
Poly(VHX)	H$_2$O/SDS[c]	0	0	0		0			0[d]	–	
Poly(VA)	H$_2$O/SDS	0	–	–		–			–	Complex Py > Pu	0
Poly(A)	H$_2$O/SDS	Complex 2U:1A	–	–		–			–	–	

[a] Combinations that were not investigated are indicated by dashes

[b] Water-propylene glycol solutions

[c] Aqueous sodium dodecyl sulfonate solutions

[d] Complexing only after heating

[e] Pu and Py denote purine and pyrimidine residues, respectively

Poly(U) = poly(uridylic acid), poly(A) = poly(adenylic acid), poly(C) = poly(cytidylic acid), poly(G) = poly(guanylic acid), poly(I) = poly(inosinic acid), poly(iA) = poly(isoadenylic acid), poly(VA) = poly(9-vinyladenine), poly(VC) = poly(1-vinylcytosine), poly(VU) = poly(9-vinyluracil), poly(VHX) = poly(9-vinylhypoxanthine)

mers have no ordered structure and showed only slight intramolecular base-base interactions, the interactions between atactic polymers are strongest. The isotactic polymers may assume a helix conformation as reported for isotactic poly(methyl methacrylate). Since nucleic acid bases are situated outside the polymer chain, they can form a complex although the interactions are relatively weak. On the other hand, the syndiotactic polymer may have a rod-like conformation. Therefore, it is well understood that the complex formation ability of syndiotactic polymers is very low. From the results of the formation of complexes containing hydrogen bonds, the following conclusions may be drawn:

(1) the formation of the complex is strongly affected by solvation of the polymer components, i.e. when increasing the dielectric constant, the complex is easily formed in aprotic solvents, whereas in protic colvents the reverse tendency is observed,

(2) in aqueous medium, there exists a critical pH when poly(carboxylic acid)s are used, i.e. cooperative interaction is involved in complexation,

(3) the effect of hydrophobic interactions on the stabilization of the hydrogen-bond containing complex may be stronger than that on the stabilization of the polyelectrolyte complex, and

(4) in heterogeneous systems, the complex formation is mainly controlled by the rate of penetration of the soluble polymers into the membrane.

3.3.2 Physical and Chemical Properties of Hydrogen-Bond Containing Complexes

Smith et al.[178] reported the unique properties of intermacromolecular PEO-PAA complexes. Figure 31 shows the temperature dependence of the stiffness (a), the dependences of the glass transition temperature (T_g) (b) and heat stability (c) on the composition of the complex, and the stress-strain curve (d). The representative properties of PEO and of its complex with PAA are compiled in Table 17. In Fig. 31(a), two transitions of the stiffness are observed for a mixture with 75 wt% PEO. The first break around 15 °C is attributed to a compatible mixture of the two polymers. The new phase, an amorphous complex, has a lower softening temperature than pure PAA. The additional polyacid has effectively eliminated the PEO crystalline phase, apparently leaving only a single and amorphous phase (50% in Fig. 31(a)). On further decrease of the PEO content, a series of similar stiffness-temperature curves approach the position of PAA. Surprisingly, however, at still lower PEO contents, e.g. 10% PEO, the stiffness values are substantially above those of each polymer component. At the same composition, the maximum of the glass transition temperature is observed (Fig. 31(b)). In Fig. 31(c), the heat stabilities are plotted against the composition of the polymer components using film samples heated in air at 300 °C for one hour. A predicted weight constancy for mechanical mixtures of the polymer components is indicated by a dotted straight line,

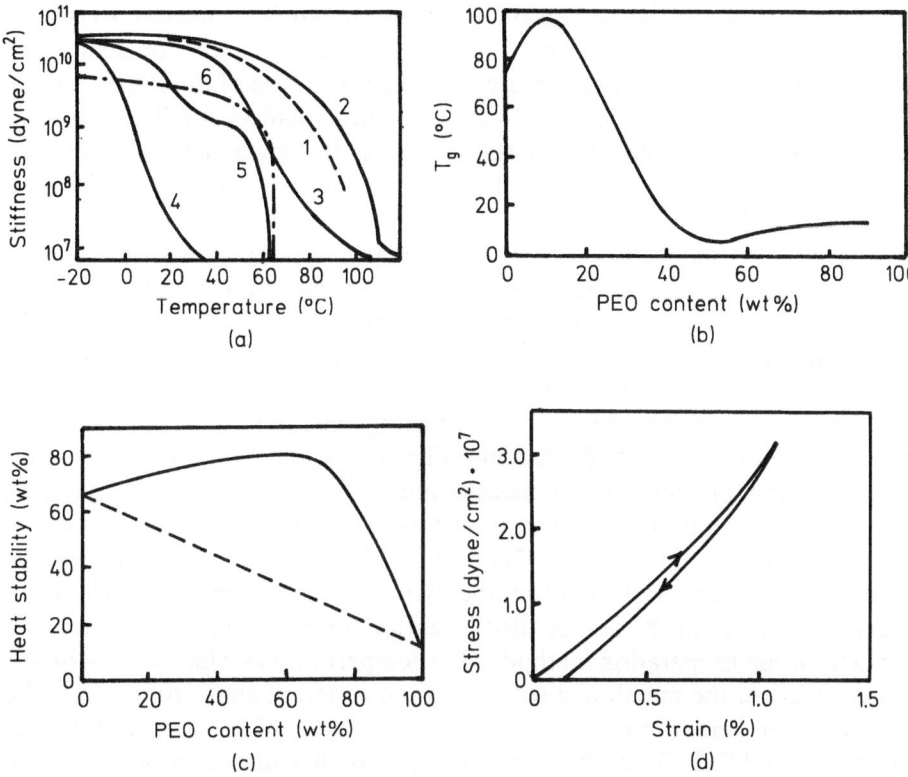

Fig. 31 a–d. Mechanical properties of poly(acrylic acid) (PAA)-poly(ethylene oxide) (PEO) complex formed by hydrogen bonds[178]. (a) Temperature dependence of stiffness; (1) PAA, (2) 10 wt% PEO, (3) 25 wt% PEO, (4) 50 wt% PEO, (5) 75 wt% PEO, (6) PEO, (b) Dependence of glass transition temperature (T_g) on the composition of PAA-PEO; (c) Dependence of heat stability on the composition of PAA-PEO retained at 300 °C for 1 h, (d) Stress-strain curve

Table 17. Characteristic properties of poly(ethylene oxide) (PEO) and its complex with poly(acrylic acid) (PAA)

Property	Complex[a]	PEO
Stiffness[b] (dyne/cm²) × 10^{-7}	1.38	210 ~ 480
Glass transition temperature (°C)	5	−55
Heat stability[c] (%)	77	5
Water extraction resistance[d] (%)	87 ~ 93	None
Shore hardness (D scale)	20	52
Tensile strength (dyne/cm²) × 10^{-7}	5.52 ~ 14.49	12.42 ~ 16.56
Ultimate elongation (%)	400 ~ 800	700 ~ 1200

[a] Complex containing 40 ~ 50 wt% PEO
[b] Secant modulus (25 °C)
[c] Wt% retention at 300 °C for 1 h
[d] Wt% of insoluble portion at room temperature

assuming that each polymer contributes proportionately (related to weight) to the heat stability of the mixtures. The upper curve shows that the actual heat stability of the complex rises rapidly with slight increase of the PAA content, reaching a maximum which levels off as a short plateau for PAA contents ranging from 30 to 45%. It is interesting that the maximum value of 79% weight constancy is higher than that of each polymer component. When a complex membrane with equal content (wt%) of the monomeric units is stretched to 1% elongation at −40 °C, a steeper slope or inverse curvature is observed (see Fig. 31(d)). Stress increases faster than strain and this probably reflects the presence of weak crosslinkages (hydrogen bonds) in the complex. Upon stretching, the complexes exhibit a higher energy of interaction or more linkages are formed, resulting in a higher degree of order.

The complex membrane pepared by mixing dimethyl sulfoxide (DMSO) or water-ethanol solutions of PAA and PVPo was studied by infrared spectroscopy and dynamic-mechanical measurements[224]. The content of hydrogen bonds of the membrane obtained in water-ethanol is higher than that of the membrane formed in DMSO. This result coincides with those obtained in solution as stated previously. From the dynamic-mechanical properties, it was found that PAA and PVPo are distributed uniformly in the membrane independent of the preparation methods. The temperature at which a peak of loss modulus (ε'') of the membrane with the composition of about 70 mol% PVPo prepared from water-ethanol occurred, was about 50 °C higher than that prepared from DMSO. This phenomenon suggests that the complex membrane formed in water-ethanol has a more ordered structure. Moreover, Tanaka et al.[224] evaluated the glass transition temperature (T_g) of this membrane which is obtained as ε'' peak temperature in comparison with T_g of copolymer or polyblend systems using the modified Gordon-Tayler equation[482];

$$T_g = \frac{w_1 \cdot K_1 \cdot T_{g1} + w_2 \cdot T_{g2}}{w_1 \cdot K_1 + w_2} - \frac{w_1 \cdot w_2 \cdot K_2}{w_1 \cdot K_1 + w_2} \tag{42}$$

where w_1 and w_2 represent the weight fractions of the polymer components 1 and 2 and T_{g1} and T_{g2} the corresponding glass transition temperatures. The first term is related to weak interactions (mixing effect) and second term to relatively strong interactions between the polymer components. Experimental T_g value may be comparatively well explained by changing K_2 (which is a parameter indicating interactions between polymers) at constant K_1 (which is a parameter indicating mainly mixing effect). On the other hand, using PGA with a specific rigid conformation (α-helix) and PEO as the components, a complex membrane of a heterogeneous network polymer was obtained[225]. This implies that the resulting polymer complexes are not simple adducts; nevertheless, PGA and/or PEO domains coexist.

Tonami et al.[179] studied atactic-PMAA (at-PMAA)-PEO complex membranes by means of infrared spectroscopy, stress relaxation measurement and torsional analysis of dynamic-mechanical properties. They pointed out that the polymer complex was formed through hydrogen bonds between the ether

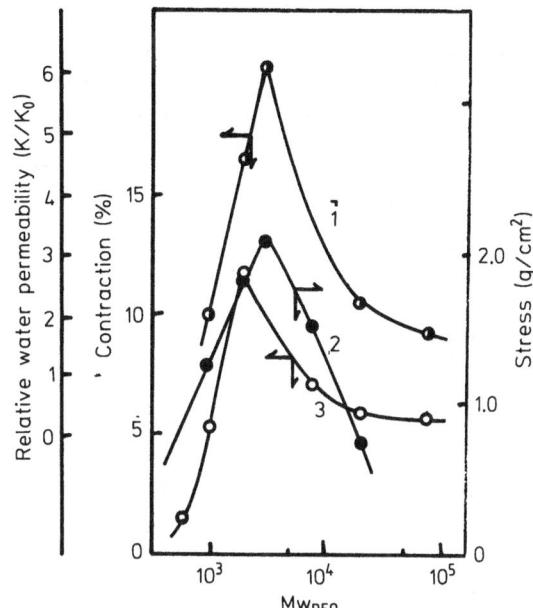

Fig. 32. Dependence of mechanochemical contraction, stress and water permeability of a complex membrane consisting of poly-(methacrylic acid) (PMAA) and poly(ethylene oxide) (PEO) on the molecular weight of PEO[483]. (1) Relative permeability to water, (2) Stress, (3) Contraction

oxygen atoms of PEO and carboxy groups of PMAA. In addition, hydrogen bonds were also detected to exist between the carboxy groups of PMAA, resulting in the formation of carboxylic acid dimers, and the energetic difference of these two types of hydrogen bonds (carboxylic acid dimer $\rightleftharpoons$ $-COOH \ldots OCH_2CH_2-$) was found to be about 4.2 kcal/mol.

There are only few data available on the permeability of intermacromolecular complex membranes formed through hydrogen bonds. However, recently a quite interesting phenomenon was reported by Osada et al.[483]. Figure 32 shows the dependence of the stress, contraction and water permeability of complex membranes composed of a PMAA membrane and PEO on the molecular weight of PEO. As noted previously, the PMAA membrane is contracted by complexation with PEO having a suitable molecular weight (≈ 2000). The water permeability reaches a maximum at $M_{PEO} = 2000 \sim 3000$. This fact suggests that the pore diameter in the membrane is controlled by complexation. This system is called "molecular valve".

Reverse osmosis properties of the membranes composed of cellulose acetate phthalate and PVPo are summarized in Table 18[484]. Table 18 reveals the rejection decreases as the values of solubility parameters (δ) of the membrane and of the solute approach one another.

3.4 Stereocomplexes

Isotactic and syndiotactic poly(methyl methacrylate) (iso-PMMA and synd-PMMA) are well-known to exhibit a strong tendency to form an intermac-

Table 18. Separation of organic solutes from aqueous solutions using cellulose acetate phthalate-poly(N-vinyl-2-pyrrolidone) complex membrane[484]

	Rejection (%)	Solubility parameter (δ_1)	Molar volume (cm³)
$CH_3CH_2CH_2OH$	26.1	11.9	74.8
$CH_3CH_2CH_2CH_2OH$	26.6	11.4	91.5
$(CH_3)_2CHCH_2OH$	32.8	11.1	92.0
$CH_3CH_2CH(CH_3)OH$	40.3	10.9	92.3
$CH_3CH_2CH_2CH_2CH_2OH$	30.9	10.9	108.8
$(CH_3)_2CHCH_2CH_2OH$	39.5	10.7	108.3
$(C_2H_5)_2CHOH$	44.0	10.5	109.4
$CH_3CH_2CH(CH_3)CH_2OH$	42.8	10.6	108.1
$CH_3CH_2C(CH_3)_2OH$	66.0	9.9	109.5
benzene	95.9	9.2	88.9
benzene–CH_3	98.1	8.96	105.7
benzene–C_2H_5	99.0	8.8	122.4
benzene (1,2–CH_3, CH_3)	99.3	9.00	120.5
benzene (1,3–CH_3, CH_3)	98.3	8.86	122.3
CH_3–benzene–CH_3 (1,4)	98.9	8.79	123.3
$(CH_3)_2CHCH_2OCOH$	84.2	8.47	116.1
$CH_3CH_2CH_2CH_2CH_2OCOCH_3$	91.9	7.37	148.9
$CH_3CH_2CH_2CH_2OCOCH_3$	74.5	8.10	131.6
$CH_3CH_2CH(CH_3)OCOCH_3$	77.5	8.10	133.3
$CH_3CH_2OCOCH_2CH_2CH_2CH_3$	74.0	8.25	132.2

50 atm., 25 ± 0.1 °C, pure water permeability: 0.0296 (cm³/cm² · h)

romolecular complex through stereospecific interactions in a suitable solvent and in bulk, which is termed "stereocomplex". This type of complex between stereoregular polymers has not yet been studied extensively, except for a few trials made with PMMA-PMAA[375, 376] and PMMA-poly(vinyl chloride) (PVC)[377, 378] systems. This complexation phenomenon has been investigated by means of techniques such as viscometry, turbidimetry, light scattering, osmometry, high-resolution and broad-line proton NMR, sedimentation in the ultracentrifuge, differential scanning calorimetry, X-ray diffractometry, and dynamic-mechanical and dielectric measurements.

3.4.1 Formation of Stereocomplexes

Stereocomplexes are obtained in suitable dilute and concentrated solutions and in bulk. The formation of the complexes is strongly affected by the kind of solvent used. The solvents can be divided into three classes[299]:
(A) strongly complexing solvents,
(B) weakly complexing solvents and
(C) non-complexing solvents (Table 19).

Figure 33 describes the relationship between the reduced viscosity and the fraction of synd-PMMA ([synd-PMMA]/([synd-PMMA] + [iso-PMMA]) with respect to their repeating units) in various solvents at 25 °C[300]. Type A solvents are polar solvents, e.g. dimethylformamide (DMF) and acetone, in which a minimum value of the viscosity of a fraction = 0.67, i.e. at an [iso]/[synd] ratio of 1/2, is obtained. In these systems, the stereocomplex tends to precipitate and the solution becomes turbid. Type B solvents include non-polar aromatic solvents such as benzene and toluene. In these solvents, an increase of the reduced viscosity compared with the theoretical values, assuming no interactions, is observed. However, this viscosity change is not stable and flattens with time, due to the formation of nearly invisible and strongly swollen gel particles. Type C solvents are chloroform and dichloromethane. In these solvents, a deviation from the reduced viscosity is not observed (i.e. a nearly additive curve indicating the absence of any association) which suggests a lacking ability for the complexation, probably because these solvents have a too high solvation power for both stereoregular PMMA. At elevated temperatures, such stereocomplexes melt within an extremely narrow temperature range (T_m). The complex crystallized from dilute solution shows melting endotherms with maxima at 206 ~ 215 °C, depending only slightly on solvent, on crystallization temperature and on composition. Figure 34 shows the effect

Table 19. Classification of solvents according to their ability of promoting the formation of stereocomplex between iso-PMMA and synd-PMMA[299]

Type A (strongly complexing)		Type B (Weakly complexing)	Type C (Non-complexing)
Dimethylformamide	Acetone	Toluene	Chloroform
Methyl isobutyrate	Acetonitrile	Benzene	Dichloromethane
Dimethyl sulfoxide	Tetrahydrofuran	Dioxane	
Tetrachloromethane	MMA monomer		

PMMA = poly(methyl methacrylate), MMA = methyl methacrylate

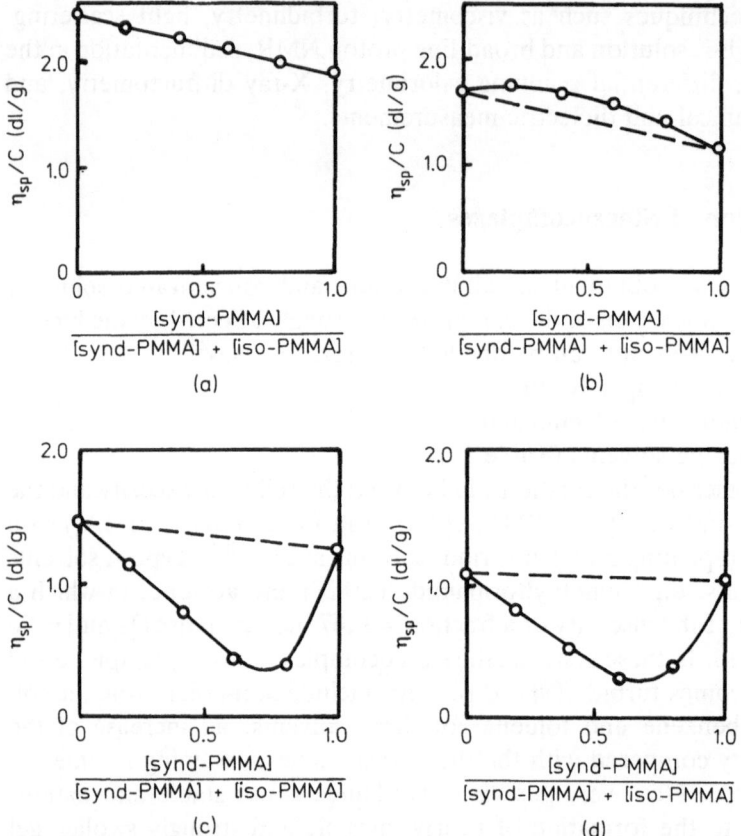

Fig. 33 a–d. Solvent effect on the formation of the stereocomplex of isotactic poly(methyl methacrylate) (iso-PMMA) with syndiotactic PMMA (synd-PMMA)[300]. (**a**) In chloroform, (**b**) in benzene, (**c**) in dimethylformamide, (**d**) in acetone

of the mixing ratio of the components of stereoregular PMMA on the melting behavior. Figure 34(a) shows that T_m is independent of the [iso]/[synd] ratio but to depend on the crystallization temperature (T_c) above 140 °C, and the equilibrium melting temperature of the stereocomplex (($T_m)_0$) lies about 260 °C ($T_m = T_c$). In the case of $T_c < 140$ °C, two neighboring melting temperatures belonging to the primary and secondary crystallization of the stereocomplex are observed. The ratios of the monomeric units, [iso-PMMA]/[synd-PMMA], of the stereocomplex have been discussed for a long time to be 1/2, 1/1 and 1/1.5. However, Challa et al.[299] found that a real stereocomplex always possessed the ratio 1/2 of monomeric units irrespective of the initial mixing ratio and the method of preparation. They studied the formation of the stereocomplex in dilute solution and its crystalline structure using various preparation methods. Figure 34(b) shows the influence of the crystallization time on the area of the melting endotherm for different mixtures. T_m hardly changes whereas the area of the melting endotherms increases with time. Especially, it should be noted that with short annealing times, a maximum

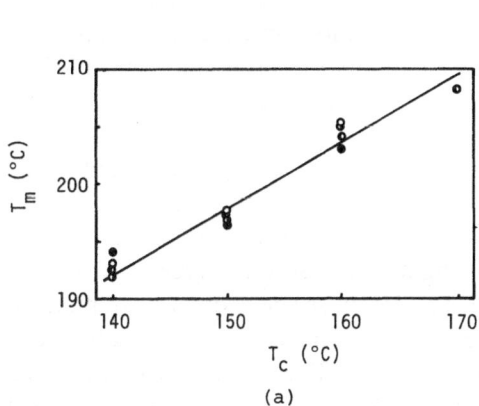
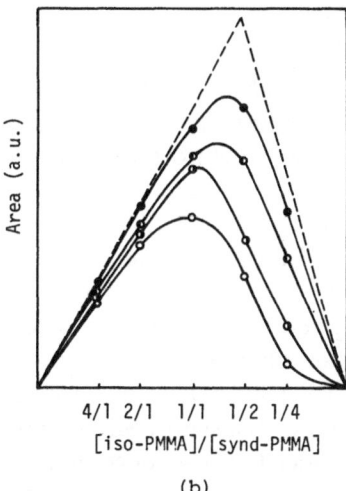

(a)   (b)

Fig. 34a, b. Effect of the mixing ratio of isotactic poly(methyl methacrylate) (iso-PMMA) and syndiotactic PMMA on the melting temperature (T_m) and specific area of the melting endotherms[299]. (a) Relationship between T_m and crystallization temperature (T_c) for different mixtures of stereoregular PMMA; [iso-PMMA]/[synd-PMMA]: ◑ 1/2, ○ 1/1, ◐ 2/1, ● 4/1, (b) Specific areas of melting endotherms of stereocomplexes with different [iso-PMMA]/[synd-PMMA] ratio in bulk mixtures. Crystallization time: ○ 4 h, ◑ 16 h, ◐ 4 days, ● 2 weeks

crystallinity is observed at [iso-PMMA]/[synd-PMMA] = 1/1, but the maximum shifts to the ratio 1/2 with increasing annealing time. The fact that an excess of iso-PMMA or synd-PMMA (deviation from specific stoichiometry = 1/2) has no effect on the melting temperature and X-ray diffraction pattern suggests that the binary interaction parameter of the stereocomplex containing excess polymer components is nearly equal to zero. Solvent effects (type A and type B) were not detected by viscometric and turbidimetric titrations and osmometry in dilute solution. From these results, it may be concluded that the stereocomplex is always formed from one part of iso-PMMA and two parts of synd-PMMA in dilute solution independent of the initial mixing ratio and the solvent (except type C solvents). However, in bulk or in extremely concentrated solution systems, since the chain mobility is depressed, the formation of the stereocomplex may take a longer annealing time, i.e. the crystallinity of the samples changes with time. Since T_m remains constant, the increase of crystallinity with annealing time is attributed to the formation of a new complex rather than to the stabilization of the existing one. This means that even in bulk and concentrated systems, the final and most suitable composition of the stereocomplex is assumed to be [iso-PMMA]/[synd-PMMA] = 1/2.

The crystalline structure of the stereocomplex has not yet been elucidated. Liquali et al.[301] proposed a model for the stereocomplex (see Eq. (13)) involving two flat syndiotactic chains with the same number of trans- and gauche-conformations which may lie in the grooves of one isotactic 5_1-helix. More recently, Tadokoro et al.[302] found that iso-PMMA forms a 10_1-helix. Furthermore, they assumed a similar double-stranded helical structure of iso-PMMA for the stereocomplex. The stereocomplex prepared by matrix polymerization (i.e. radical polymerization of methyl methacrylate in the presence of stereoregular (usually syndiotactic) PMMA[303–305]) exhibits a more compact and complete structure than that prepared by mixing iso-PMMA and synd-PMMA. This is confirmed by the elevated T_m, suggesting that the growing chain may be more or less parallel to the matrix chain.

Liquali et al. assumed that the α-methyl groups play the dominant role in the formation of the stereocomplex[301]. However, Spěváček et al.[306] investigated the great importance of the ester groups. They determined the relaxation time (T_2) of the stereocomplex using a high-resolution (HR-), broad line (BL-) and magic angle rotation (MAR-) NMR pulse spectrometer. In mixtures of iso- and synd-PMMA in benzene, the system behaves as a network linked by associated segments of iso- and synd-PMMA. These junctions (associated stereocomplex) are connected by non-associated segments. In mixtures of iso- and synd-PMMA in CD_3CN or CCl_4, almost all monomeric units are associated. In the associated stereocomplex itself there exist two types of protons differing in mobility. From the MAR-NMR analysis (Fig. 35), it can be concluded that in the PMMA gels in benzene the more mobile protons correspond to α-CH_3 and CH_2 groups whereas the ester groups are practically immobilized. Stereo-association of iso-PMMA and synd-PMMA in benzene is considered to be affected by the interaction of ester groups. From these results it may be suggested that even in non-aromatic solvents such as DMF, acetonitrile and CCl_4, the stereocomplex is also formed mainly through the interaction between ester groups.

Similar stereospecific associations have been investigated in the systems iso-PMMA-syndiotactic-PMAA (synd-PMAA)[375, 376] and synd-PMMA-PVC[377, 378]. Figure 36 shows the reduced viscosity as a function of the molar

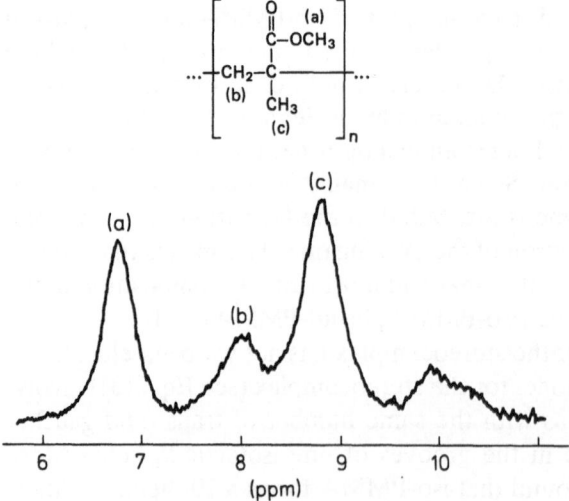

Fig. 35. Magic angle-rotation NMR spectrum of the stereocomplex of stereoregular poly(methyl methacrylate) (PMMA) in C_6D_6[306]. [iso-PMMA]/[synd-PMMA] = 1/2

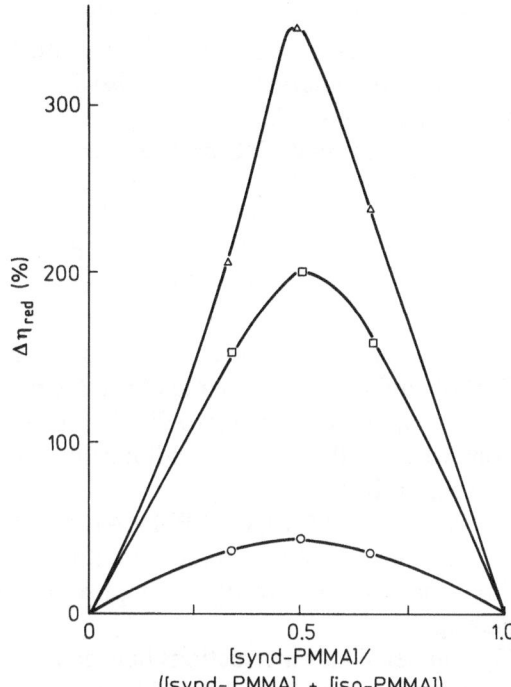

Fig. 36. Composition of the stereocomplex in the system of isotactic poly-(methyl methacrylate) (iso-PMAA) and syndiotactic PMMA (synd-PMMA)[376]. Time after mixing: ○ 1 h, □ 4 h, △ 8 h

fraction of the synd-PMAA (χ_s) unit in the iso-PMMA-synd-PMAA system. The maximum of the reduced viscosity at $\chi_s = 0.5$ suggests the formation of a stereocomplex with an equimolar composition. T_m^s of the iso-PMMA-synd-PMAA complex is 61 ~ 63 °C in DMF which is lower than that of iso-PMMA-synd-PMMA. This means that the interaction between iso-PMMA and synd-PMAA is weaker than that between iso- and synd-PMMA. The solvents used are divided into three classes (cf. Table 19). A pure solvent of type A has not yet been found but a mixture of 83 vol% of ethanol with water at 45 °C belongs to this type of solvent promoting the formation of an associate with the ratio [iso-PMMA]/[synd-PMAA] = 1/2. Type B solvents are further classified into two subclasses; the first subclass includes DMF, N,N-dimethylacetamide (DMA) and DMSO in which the viscosity rises continuously over many hours to high reduced viscosity values. The second subclass includes diethylene glycol monoalkyl ethers, e.g. ethylene glycol monomethyl ether and 2-methoxyethanol, in which the viscosity reaches a rather low maximum within a relatively short period of time. In type B solvents, the most suitable ratio of the monomeric units of the complex appears to be 1/1 (see Fig. 36). A type C solvent is pyridine. Moreover, the importance of the interactions between the ester groups of iso-PMMA and the α-methyl groups of synd-PMAA (not hydrogen bonds) has been studied on the basis of the experiments of association between stereoregular PMMA and partially hydrolyzed stereoregular PMMA (i.e. methyl methacrylate-methacrylic acid copolymers). Substitution of about 15% of MMA by MAA in iso-PMMA is sufficient to suppress association completely. It is quite interesting that of the four possible cross combinations between tactically regulated PMMA and PMAA (isotactic and syndiotactic), only the combination between iso-PMMA and synd-PMAA can form the stereospecific complex.

In conventional PVC and stereoregular PMMA systems, it was found that iso-PMMA and PVC form an incompatible system over the entire composition range whereas synd-PMMA and PVC form a compatible system up to a ratio of [PMMA]/[PVC] $\simeq$ 1. This system dissolves in excess PVC yielding only one value for T_g but is insoluble in excess synd-PMMA. The ester groups of PMMA are considered to act as proton-accepting groups, and PVC behaves like a weak proton-donating polymer due to its α-H atoms[485]. Therefore, PMMA and PVC are capable of producing a stereocomplex, i.e. they exhibit a good compatibility, owing to dipole-dipole and/or hydrogen bond-like interactions when PMMA has a syndiotactic structure.

To summarize:

(1) the [iso-PMMA]/[synd-PMMA] ratio of the stereocomplex is 1/2,

(2) the solvent effects on the formation of the stereocomplex are divided into three classes, i.e. strongly complexing, weakly complexing and non-complexing,

(3) van der Waals interactions between the ester groups of iso-PMMA and the α-methyl groups of synd-PMMA are important, and

(4) there are further stereospecific associations.

3.4.2 Physical and Chemical Properties of Stereocomplexes

Samples of the stereocomplexes of iso-PMMA and synd-PMMA for dynamic-viscoelastic measurements can be prepared in different ways, i.e. as a solid from type A solvents[307, 308], as a gel from type B solvents[309, 310] and by matrix polymerization[311]. Differential scanning calorimetry (DSC) of the solid stereocomplex revealed that endothermic peaks are located at 185 °C, corresponding to the melting point of the induced crystalline part of the free synd-PMMA (purified synd-PMMA does not crystallize under the same condition), at 210 °C, corresponding to the melting point of the stereocomplex and at 280 °C, corresponding to the depolymerization temperature. The degree of crystallization of the stereocomplex was assumed to be about 5 to 40% by X-ray diffraction. When the stereocomplex was prepared in bulk instead of by precipitation from type A solution as stated above, the same thermal properties were observed only by annealing of the complex at 240 °C for more than 2 min and subsequent rapid cooling to 150 °C. This result suggested that the stereocomplex was not varied by the different preparation methods[319]. In Fig. 37, the dynamical-mechanical behavior (loss tangent, tan δ) of the PMMA stereocomplex is described as a function of the ratio of the monomeric units, [iso-PMMA]/[synd-PMMA]. At an annealing temperature of 140 °C, crystallization was completed after 40 h. However, the tan δ peak is shifted

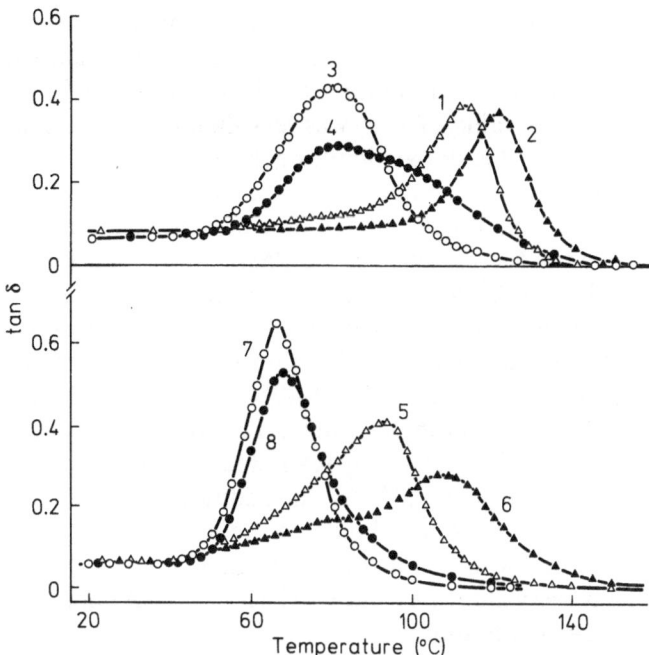

Fig. 37. Effect of annealing on loss tangent (tan δ) of mixtures of stereoregular poly(methyl methacrylate) (PMMA)[307]. (1) [iso-PMMA]/[synd-PMMA] = 1/4, before annealing; (2) [iso-PMMA]/[synd-PMMA] = 1/4, after annealing at 140° for 120 h. (3) [iso-PMMA]/[synd-PMMA] = 3/2, before annealing; (4) [iso-PMMA]/[synd-PMMA] = 3/2, after annealing at 140° for 120 h. (5) [iso-PMMA]/[synd-PMMA] = 2/3, before annealing; (6) [iso-PMMA]/[synd-PMMA] = 2/3, after annealing at 140° for 120 h. (7) [iso-PMMA]/[synd-PMMA] = 4/1, before annealing; (8) [iso-PMMA]/[synd-PMMA] = 4/1, after annealing at 140° for 120 h

toward higher temperatures with additional broadening as shown in Fig. 37[319]. This shift may be partly due to stresses created in the amorphous region as a result of the formation of tie molecules between the complex crystallites and to the increase of the content of synd-PMMA in the amorphous region. The storage modulus (ε') above the glass transition temperature was not affected by annealing. On the basis of this result, it may be assumed that the main process taking place in bulk during annealing is the growth of complex crystallites without changing cross-linking, i.e. the number of microcrystallites.

A secondary stereocomplex gel with the ratio [iso-PMMA]/[synd-PMMA] = 1/1 obtained from o-xylene solution showed an endothermic peak at 110 °C which was attributed to the melting point of the stereocomplex. The degree of crystallization was about 5–6%. Since this gel contained over 80% solvent, micro-Brownian motions of PMMA chains in the amorphous region were relatively unhindered. When the PMMA gel was cooled rapidly after melting above its T_m (i.e. above 145 °C), first the loss modulus (ε'') changed relative to the viscosity of the system whereas the storage modulus began to increase after some time. This result confirms the assumption that first the polymer chains cannot be cross-linked but are weakly associated with one other, and then a network structure is generated with the growing of the crystalline portion after the formation of nuclei. ε' and ε'' reached constant values within 2 to 20 days. ε' was varied from about 10^{-1} dyne/cm^2 at 100 °C to about 10^6 dyne/cm^2 at 30 °C. The value of $\varepsilon''/\varepsilon'$ was less than 0.05 with a large elastic contribution. Therefore, this stereocomplex gel shows almost the same dynamic-mechanical properties as a chemically cross-linked swollen gel.

When matrix polymerization of methyl methacrylate monomer was performed in the presence of iso-PMMA or of the stereocomplex ([iso-PMMA]/[synd-PMMA] = 1/1), the tan δ peaks of the α-process (caused by the micro-Brownian motions of main chain) and the β-process (caused by the motions of the ester side chains of PMMA) shifted to temperature ranges which were higher by 5–10 °C and 30–40 °C, respectively, as compared with synd-PMMA[312]. From these results, Tanzawa et al. concluded that the stereospecific association drastically restricted the motion of the ester side chains of the two stereoregular PMMAs. In addition, the mutual interlocking of the side chains of PMMA also hindered, to a certain extent, the motion of the main chains of PMMA. Furthermore, they studied the matrix polymerization of methyl acrylate (MA) and ethyl acrylate (EA) in the presence of stereoregular PMMA and found:

(1) the restrictions imposed on the motions of the flexible MA and EA chains by the rigid PMMA chains increased in the following order: blended polymer system, stereocomplex system, and copolymer systems of MA or EA and MMA

(2) the compatibility of the MA chain with the PMMA chain was higher than that with the EA chain. These facts demonstrate that EA or MA chains and PMMA chains are distributed in the stereocomplex more homogeneously than in the blended polymers, but more heterogeneously than in the copolymers. Therefore, although being a kind of blended polymers, the stereocomplex exhibits a transparency like the copolymer and mechanical properties different from those of the usual blended polymers and copolymers.

The optical anisotropy of the solution of the stereocomplex, which was obtained by matrix polymerization[486], was characterized by the reduced anisotropy $[\Delta n/g(\eta - \eta_0)]_{g \to 0}$ where Δn denotes the flow birefringence, g the velocity gradient, and η and η_0 the viscosities of the solution and solvent[313]. It is obvious that the presence of even a small amount of synd-PMMA in the system, in which polymerization takes place, leads to an increase in the degree of regularity of the orientation of the macromolecules. A maximum is observed at a 1/2 ratio of synd-PMMA to formed polymer. The reduced optical anisotropy values for synd-PMMA are positive. Thus, when no interactions are assumed to occur between new propagation chains and matrix synd-PMMA chains, the absolute value of the reduced optical anisotropy should to be lower whereas the experiments yield a higher value. Moreover, the optical anisotropy of solutions of stereocomplexes generated during the polymerization process was found to be much lower than that for the PMMA stereocomplex prepared by mixing synd- and iso-PMMA[314]; for the latter stereocomplex, the maximum degree of anisotropy was observed at a ratio of 1/1. This difference may be attributed either to a less compact packing of the stereocomplex structures formed during polymerization or to a lower content of isotactic diads in the macromolecules formed on the matrix of synd-PMMA chains.

The permeabilities of the stereocomplex membranes to water and NaCl are shown in Fig. 38 as a function of the water content or the composition (isotactic mol%). The permeability ratios of urea to NaCl are summarized in Table 20[315]. The dependence of water permeability on the water content is very strong whereas that of the permeability to solutes (in this case NaCl) is weak. The water permeability of the PMMA membrane is extremely high compared to that of cuprophan whereas its permeability to solute is almost the same. This characteristic permeability accounts for the effect of the hyd-

Table 20. Ratios of urea permeability to NaCl permeability of stereocomplex membranes[315]

Water content (wt%)	$\dfrac{P_2 \text{ (urea)}}{P_2 \text{ (NaCl)}}$
43	1.0
57	1.0
68	1.1
74	0.9

Ratio of diffusion coefficient of urea to NaCl; D(urea)/D(NaCl) = 0.86

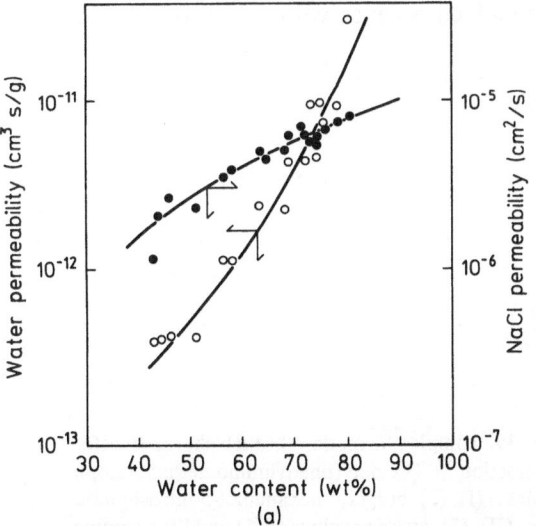
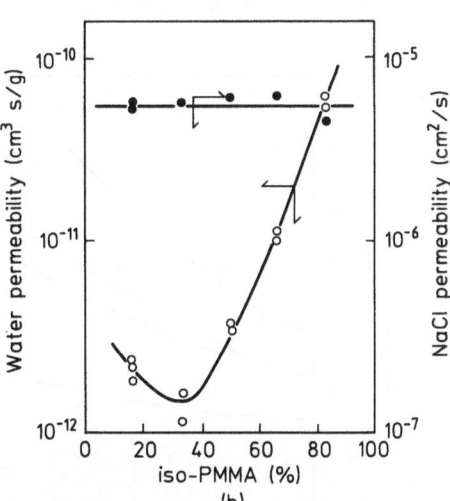

Fig. 38 a, b. Permeabilities of stereocomplex membranes to water and NaCl[315]. (a) Effect of water content, (b) effect of isotactic poly(methyl methacrylate) content

rophobicity of the membrane materials. The ratios of urea permeability to NaCl permeability for various water contents of the membranes are constant over a wide range of water content. These ratios are almost equal to the ratios of the diffusion coefficients of urea to NaCl. This result suggests that there is no characteristic interaction between the membrane material and the solute molecules because of the simple chemical structure of the membrane materials. Moreover, the water permeability reaches a minimum value at a content of about 30% iso-PMMA, i.e. the finest membrane structure is obtained, and the solute permeability remains almost constant throughout the experiments. Furthermore, the permeability of this membrane is explained by a capillary model allowing for slight tortuosity. From these results it is concluded that there is a close relationship between complex formation and membrane permeability. The antipathic behavior of synd-PMMA-conventional PMMA stereocomplex systems was also studied[316].

3.5 Charge-Transfer Complexes

The importance of charge-transfer complexes (C-T complexes), most of them being composed of small molecule-small molecule or small molecule-macromolecule, has been pointed out in various fields, e.g. photochemistry[487], electrochemistry[488] and biochemistry[489]. The systematization of the procedures for the preparation of C-T complexes in low molecular weight compound systems has also been reported. However, there are only few studies on C-T complexes in intermacromolecular complex systems available, because electron-accepting polymers are difficult to synthesize.

First, Sulzberg et al.[381, 382] studied the synthesis of high molecular weight electron-accepting polymers and their C-T complexes with electron-donating

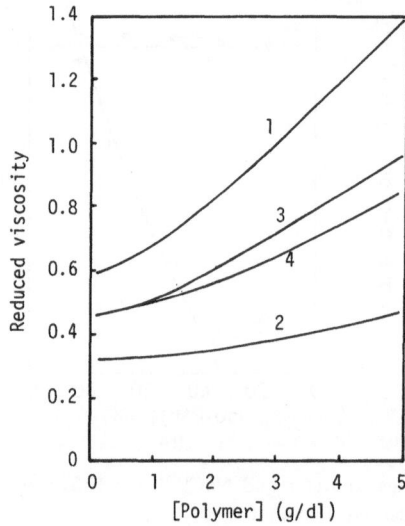

Fig. 39. complex formation through charge-transfer interaction[382]. (1) poly(phenyliminodiethanol isophthalate) (I), (2) poly(1,6-hexanediol-5-nitrosophthalate) (II), (3) average values of (1) and (2) assuming no interactions, (4) experimental values of a mixture of polymers (I) and (II)

Table 21. Properties of mixtures of donor and acceptor polymers $(1:1)^{a\ 381)}$

Property[b]	(I) + (IV)	(II) + (V)	(III) + (VI)
Color	Yellow	Yellow-orange	Yellow-orange
Lowest energy absorption maximum (nm)	391	398	398
Reduced viscosity[c] (ml/g)	0.50	0.31	0.60
Glass transition temp. T_g (°C)	50	90	60–70
Tensile modulus (psi)	2.6×10^5	2.35×10^5	3.6×10^5
Volume resistivity (ohm · cm)	1×10^{13}	–	–

[a] This mixture was prepared from 10% solutions (by weight) of donor polymer and acceptor polymer
[b] Films prepared by casting chloroform solution
[c] Determined in chloroform at 25 °C at 0.2 g/100 ml
Donor polymers; (I) = poly(phenyliminodiethanolisophthalate), (II) = poly(p-anisylimino-diethanolisophthalate), (III) = poly(p-anisyliminodiethanolbisphenol A carbonate)
Acceptor polymers; (IV) = poly(1,6-hexanediol-5-nitroisophthalate), (V) = poly(bisphenol A 5-nitroisophthalate), (VI) = poly[bis(2-hydroxyethyl)-5-nitroisophthalate-bisphenol A carbonate]

polymers (see Eq. (15)). As shown in Fig. 39, the experimentally determined values of the reduced viscosity of a mixture of an acceptor polymer and a donor polymer deviate from the average values calculated from the data of each polymer assuming no interaction. This result demonstrates that chain-to-chain interactions between both polymers may occur probably due to charge-transfer interactions. Table 21 shows the physical properties and electrical conductivities of some C-T complexes in intermacromolecular systems. The physical properties of the complex slightly deviate from those of each polymer component. In addition, the volume resistivity of the complex is increased by about three orders of magnitude compared with the donor polymer component.

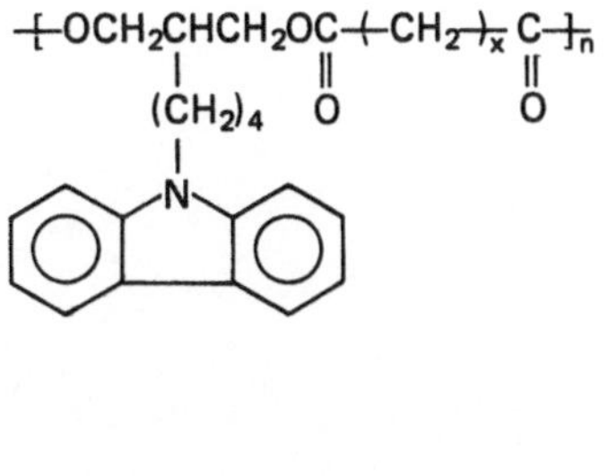

Electron–donating polymer

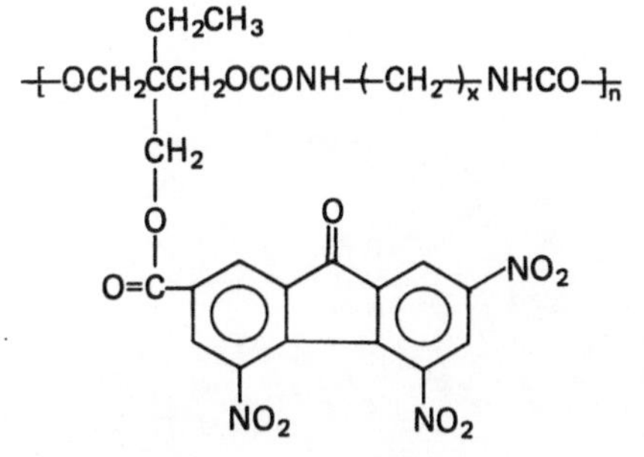

Electron–accepting polymer

Tazuke et al.[592] reported the intermacromolecular C-T complex of a poly-ester bearing pendant carbazolyl groups (electron-donating polymer) with polyurethanes bearing pendant trinitrofluorenonyl groups (electron-accepting polymer).

They suggested that the formation of the complex depended on the structure as well as on the degree of polymerization. When the intervals of active sites are identical, tight complex was formed. However, the stoichiometry of the complex was not clarified.

To develop further C-T complexes composed of macromolecules, it is necessary to synthesize new electron-accepting polymers with good physical and chemical properties but up to now, only a few syntheses of such polymers have been reported[490–494]. C-T complexes offer new application possibilities of polymers as high-conductive, heat-resistant, energy-transferrable, and charge-separable materials.

4 Cooperativity of Complexation Between Macromolecules

Though a macromolecular chain is composed of an accumulation of certain repeating units, its characteristics and reactivity are not considered to be simply the sum of the repeating units. Rather, the characteristics and reactivities are attributable to interactions between repeating units. Therefore, one cannot discuss polymer effects without taking into account inter- and intrachain interactions. However, only few data on these types of interactions between polymers have been reported. The polymer effects are characterized as follows:

(1) cooperative interactions,
(2) concerted interactions,
(3) complementarity of structures,
(4) appearance of specific microdomains.

These polymer effects account for the reaction specificity and substitution reactions between macromolecular chains.

4.1 Cooperative Phenomena in Complexation Processes

It has been stated that many kinds of interaction forces act concertedly in the formation of intermacromolecular complexes. Thus, the free energy change of the reaction between two different macromolecular chains (ΔF^0) is represented by the following equation:

$$\Delta F^0 = \Delta f_C^0 + \Delta f_{hb}^0 + \Delta f_{hp}^0 + \Delta f_v^0 + \ldots. \tag{43}$$

where Δf_C^0, Δf_{hb}^0, Δf_{hp}^0, and Δf_v^0 represent the free energy changes caused by Coulomb interactions, hydrogen bonding, hydrophobic interactions, and van der Waals forces, respectively. Even if different binding sites, which can react with complementary ones through the same interaction forces, coexist within one macromolecular chain, are these interactions considered to act concertedly. In this review, cooperative interactions are defined as interactions between complementary macromolecules when each macromolecule is composed of only one kind of repeating unit. In physical chemistry, "cooperativity" refers to a rather wide range of systems and phenomena, e.g. melting of crystals and crystallization, conversion of paramagnets to ferromagnets, denaturation and renaturation of proteins, and helix-coil transitions of poly(amino acid)s. In these systems, the state and behavior of every unit essentially depend on its ineractions with neighboring units.

Polymers are generally known to have a tendency to form intermacromolecular assemblies, even in dilute solutions as previously stated. This is attributed to the cooperative character of the intermolecular linkages, due to the long-chain structure of the macromolecules. In a system containing, for example, a pair of complementary macromolecules, the energy for breaking one bond of the inner repeating unit is not compensated by an increase in entropy, because this process is not accompanied by additional and translational degrees of freedom, which are, however, the case if the bond between two corresponding monomers is broken. In other words, cooperative stabilization may exist owing to the entropy factor. Figure 40 schematically describes the change in thermodynamic parameters through the formation of, for example, polyelectrolyte complexes. If one of the active sites reacts with a complementary one, the neighboring active sites have a more favorable entropy to form the new bindings. Thus, the activation entropy ($\Delta S^{\neq}$) is highest for the first binding and becomes lower suddenly according to the increase of the number of binding sites. Therefore, the reaction between two macromolecular chains advances very rapidly like a zipping mechanism. On the other hand, the thermodynamic parameters of the complex formation between macromolecular chains, which is directly connected with the stability of the intermacromolecular complexes, are as follows. The enthalpy change (Δh_i) is not very large and only slightly depends on the number of active sites, because the formation of the polyelectrolyte complexes may be regarded as a kind of substitution reaction of counterions. Naturally, in aqueous medium, the hydrophobic interactions also acts concertedly. Since, however, hydrophobic

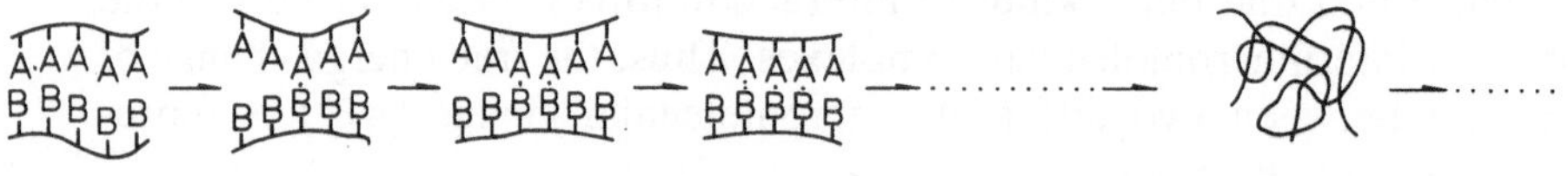

ΔS_1^*	$>>$	ΔS_2^*	$>$	ΔS_3^*	$>$		$>$	ΔS_i^*	$<$	ΔS_{i+1}^*	$(\simeq \Delta S_n^*)$
Δs_1	$<$	Δs_2	$<$	Δs_3	$<$		$<$	Δs_i	$>$	Δs_{i+1}	$(\simeq \Delta s_n)$
Δh_1	$\simeq$	Δh_2	$\simeq$	Δh_3	$\simeq$		$\simeq$	Δh_i	$\simeq$	Δh_{i+1}	$(\simeq \Delta h_n)$
Δf_1	$>$	Δf_2	$>$	Δf_3	$>$		$>$	Δf_i	$<$	Δf_{i+1}	$(\simeq \Delta f_n \simeq 0)$
k_1	$<$	k_2	$<$	k_3	$<$		$\simeq$	k_i	$\simeq$	k_{i+1}	$(\simeq k_n)$

Fig. 40. Schematic representation of the cooperative interaction between two complementary macromolecular chains

interaction forces mainly result from a change of entropy, their contribution to the enthalpy change, Δh_i, is negligible. On the other hand, the entropy change (ΔS_i) is considered as the sum of the entropy change caused by the fixation of polymer chains (ΔS_p), the entropy change caused by the released microions (ΔS_m) and the entropy change resulting from hydrophobic interactions (ΔS_h). When the increase in ΔS_h and ΔS_m exceeds the decrease in ΔS_p, the reaction between macromolecular chains is favorable with respect to entropy and advanced. ΔS_m is not changed to the some extent by the number of active sites whereas ΔS_h increases and ΔS_p decreases with growing number of active sites, i.e. with the chain length of the component polymer. Therefore, the total entropy change in the formation of intermacromolecular complexes gradually increases with rising chain length of the polymer components. The free energy change ($\Delta f_i = \Delta h_i - T\Delta S_i$) first decreases with growing number of bindings and then increases to about zero because of the effect of entropy-enthalpy compensation[495]. As a result, the total free energy change (ΔF) of the complex formation may gradually decrease with increasing chain length of the polymer component.

$$\Delta F = \sum_i \Delta f_i = \sum_i (\Delta h_i - T\Delta S_i) \tag{44}$$

Even relatively weak van der Waals and dipole-dipole interactions very often prove sufficient for maintaining the stabilization of some macromolecular associates in solution[496]. This may be exemplified by the stereocomplexes or the associates of the right- or left-handed helices of poly (γ-methyl-L(or -D)-glutamate) in organic solvents[497].

The theoretical treatment of cooperativity in systems composed of oligomer and polymer has been discussed in detail by Kabanov et al.[498]. The change in free energy (ΔF) in such systems is expressed by

$$\Delta F = \Delta F_1 + \Delta F_2 \tag{45}$$

$$\Delta F_2 = -T\Delta S_2 \tag{46}$$

$$\Delta F_1 = \Delta H_1 - T\Delta S_1 \tag{47}$$

where ΔS_2 means the change in configurational entropy of the system when oligomer chains are bound to a polymer matrix, and ΔH_1 and ΔS_1 are the changes in enthalpy and entropy caused by variations in the degrees of freedom of the matrix polymer, oligomer and solvent. For simplicity, they made the following assumptions:
(1) the chain length of the polymer is much longer than that of the oligomer
(2) the mole numbers of repeating units of the polymer and oligomer are the same in the complex
(3) long range interactions and hydrophobic interactions are disregarded in analytical calculations

(4) the contribution of all the interactions to ΔF is proportional to the total quantity of the complex, i.e. to the mole number of the repeating units of the oligomer n_{ol}^{pc} or of the polymer matrix n_m^{pc} present in the complex. This contribution is independent of the chain length of the oligomer or the degree of filling (β) of the matrix by the oligomer; $\beta = n_{ol}^{pc}/n_m^0 = n_m^{pc}/n_m^0$, where n_m^0 is the total mole number of the repeating units of the polymer,

(5) ΔF_1^0 (see Eq. (48)) is independent of the chain length of the oligomer,

(6) the oligomer is present in only two states, namely free or bound to the matrix, i.e. the number of bonds in the complex is close to the number of the oligomer units (ν). In accordance with this assumptions, Eq. (47) is converted into

$$\Delta F_1 = n_{ol}^{pc} \cdot \Delta F_1^0 = n_{ol}^{pc} (\Delta H_1^0 - T\Delta S_1^0) \tag{48}$$

where ΔF_1^0, ΔH_1^0 and ΔS_1^0 are the thermodynamic changes when one monomer unit of the complex is formed without making allowance for the change in the configurational entropy of the system. In the theory of complex formation equilibria[499, 500], the equilibrium constant K_i (formation of the i-th bond following after $(i - 1)$ bond; $K_i = \exp(-\Delta F_i/kT)$) is the same when the assumption (4) is valid, for example for systems of complexes formed through relatively weak interactions such as hydrogen bonds and van der Waals interactions. However, K_i (and hence ΔF_i) depends on i, when weak polyelectrolytes (the matrix or the oligomer) participate in the complexation and become charged through this process[501-503] (see Fig. 40). On the other hand, the quantity ΔS_2 is related only to the change in the thermodynamic probability of the oligomer of being in solution or on the matrix. Then, the overall equilibrium constant is simplified, if $\nu \gg 1$ and $\nu(1 - \beta)/\beta \gg 1$,

$$\frac{\beta}{(1 - \beta)m_{ol}} \exp \frac{\beta}{1 - \beta} = K_\nu \tag{49}$$

$$K_\nu = K_1^\nu = \exp(-\nu\Delta F_1^0/RT) \tag{50}$$

where K_ν is the equilibrium constant for the combination of the matrix polymer with an oligomer consisting of ν units, m_{ol} is the molar fraction of monomer units of free oligomer in solution, and $K_1 = \exp(-\Delta F_1^0/RT)$.

On the basis of these theroretical considerations, the following phenomena are expected to occur in oligomer-polymer reactions:

(1) the overall equilibrium constant is strongly dependent on the number of active sites (chain length) of the oligomer and increases rapidly when a certain degree of polymerization of the oligomer is reached (critical chain length),

(2) even weak interactions can form stable complexes when the chain length of the oligomer is longer than the critical chain length of the oligomer, and

(3) the stable complex can be formed even at very low absolute values of ΔF_1^0, because, in addition to the main interactions between complementary binding

sites, other weak interactions may contribute substantially to the stabilization of the complex.

To demonstrate such cooperative interactions experimentally, the relationship between the stability constant of the polyelectrolyte complex and the chain length of the polyelectrolyte components in the system of PMAA and quarternized oligoethylenimines with different chain length was studied[124]. As shown in Fig. 41, the stability constant of the complex increases exponentially and the free energy change in the complex formation decreases almost linearly when the degree of polymerization of oligocations is below 4. These relationships are described by the equations,

$$K = A \cdot e^{Bn} \tag{51}$$

$$-\Delta F^0 = \alpha n + \beta \tag{52}$$

In Eqs. (51) and (52), α is comparable with a cooperative coefficient, β representing the basic bonding constant. The values of A, B, α and β are complied in Table 22. In the range of the chain lengths of these oligocations, the standard free energy change of complex formation decreases by about 0.6 kcal/mol upon the addition of one cationic residue. The difference of β (about 1.6 kcal/mol) between two different types of cations may be due to the variation of the hydrophobicities between benzyl and methyl residues.

Thus, when the total free energy change exceeds the kinetic energies of the polymer chain components, a stable intermacromolecular complex is formed.

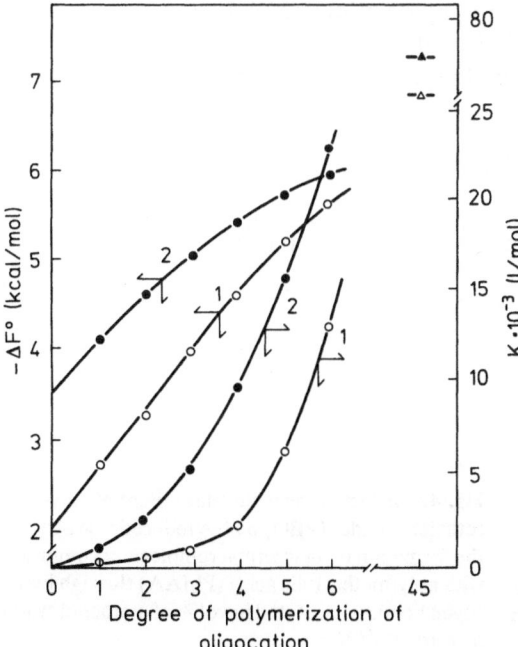

$$H(CH_2-\underset{\underset{CH_3}{|}}{\overset{\overset{R}{|}}{N^+}}-CH_2)_n H \quad I^-$$

Fig. 41. Dependence of the stability constants (K) and the free energy changes ($-\Delta F^0$) in the formation of polyelectrolyte complexes on the degree of polymerizaton of oligocations in poly-(methacrylic acid) (PMAA; degree of neutralization = 1.0, $M_w = 5.3 \times 10^4$)-oligocation systems. Oligocation; (1) R = CH_3, (2) R = $CH_2C_6H_5$

Table 22. Complex formation parameters in PMAA-oligocation systems[124)]

Cations	A	B	α	β
(I)	31.9	1.0	0.63	2.05
(II)	49.0	0.8	0.47	3.66

For A, B, α, and β see Eqs. (51) and (52); for oligocations (I) and (II) see Fig. 41

When using low molecular weight compounds whose degree of polymerization is below four, the stable complex is not formed. The critical chain length for the formation of an intermacromolecular complex is defined as the minimum chain length to form the stable complex. Figure 42 shows the relationship between the molecular weight of PEO and the change in the reduced viscosity of mixed solutions of PMAA-PEO[189)]. The viscosity is remarkable reduced due to the formation of the stable complex, because the polymer chains are contracted by desolvation. Figure 42 shows that the reduced viscosity only slightly changes when PEO with low molecular weight units (below a certain molecular weight) is used but drastically decreases above a molecular weight of ca. 2000 of PEO in aqueous medium. Thus, the critical chain length of the resulting intermacromolecular complex is about 40 as the degree of polymerization of PEO. The critical chain lengths of various intermacromolecular complex systems are summarized in Table 23. Under these experimental conditions, the molecular weight of poly(carboxylic acid)s is sufficient for the formation of a stable complex through Coulomb forces or hydrogen bonds. In polyelectrolyte complex systems, the average degree of polymerization of oligocations is four. In comparison with quaternized oligoethylenimine (QOEI), the critical chain length of OEI is longer because OEI is difficult to protonate completely. As Coulomb forces are stronger than hydrogen bonds, the critical chain length of

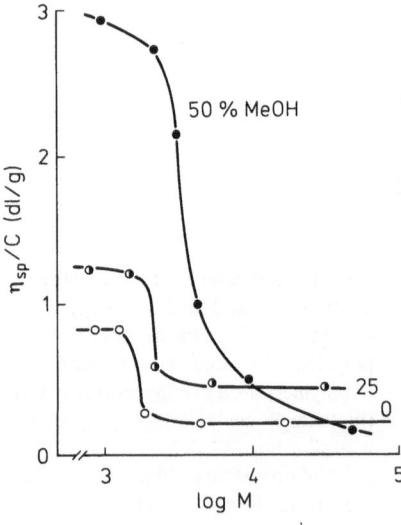

Fig. 42. Effect of the molecular weight of poly-(ethylene oxide) (PEO) on the reduced viscosity in the formation of an intermacromolecular complex with poly(methacrylic acid) (PMAA) through hydrogen bonds (designated in vol% of methanol-water mixture at 25 °C)

Table 23. Critical chain length of various intermacromolecular complexes

System	Solvent	$\overline{Pn}^a$	Main interaction force
PMAA-QOEI	Water	4	
PAA-QOEI	Water	5	Coulombic force
PGA-QOEI	Water	4	
PGA-OEI	Water	5	
PGA-OEI	Water-MeOH	6	
PMAA-PEO	Water	40	
PAA-PEO	Water	200	
PMAA-PEO	Water-MeOH	120	Hydrogen bond
PAA-PEO	Water-MeOH	200	

[a] $\overline{P}_n$ = degree of polymerization of QOEI, OEI and PEO
    PMAA = poly(methacrylic acid), PAA = poly(acrylic acid), PGA = poly(L-glutamic acid),
    OEI = oligo(ethylenimine), QOEI = Quaternized OEI (see Fig. 41) PEO = poly(ethylene
    oxide)

the complex formed through Coulomb forces is much shorter than that formed through hydrogen bonds. In the hydrogen bonds containing complex system, the critical chain length expressed as the degree of polymerization of PEO is about 40 for PMAA and about 200 for PAA in water, respectively. This difference of the critical chain length is partly explained by the difference in hydrophobic interactions. The critical chain length of the PAA system is independent of the methanol content of the water-methanol mixture whereas that of the PMAA system increases with rising methanol content. This fact also indicates the importance of hydrophobic interactions for the complexation. Thus, methanol is considered to destroy the hydrophobic interactions. Hence, the total interaction forces between PMAA and PEO are weakened by the addition of methanol.

When a polydispersed oligomer is used, the distribution of bound and free oligomers is determined by a computer simulation[504]. The addition of matrix polymer to the oligomer solution causes the molecular weight distribution of the free oligomer to become narrower, as a result of the rapid decrease of higher molecular weight fractions of the oligomer in solution which are bond selectively on the matrix, while the proportion of the lower molecular weight fractions remain practically constant in solution. Moreover, these findings were checked by the experimental results obtained from the PVPo-PAA system[504].

If the oligomer interacts with the matrix polymer cooperatively, the oligomer chains bound on the matrix interact with each other, because the free part (not covered by the oligomer) of the matrix and the bound oligomer is affected by whether neighboring sites of the matrix are covered with other oligomer chains or not. If this applies, a substantial deviation of the distribution of the oligomer chains bound on the matrix from a random distribution, in other word an "all-or-none"-type complexation, may take place (see Eq. 23). If the matrix chains are flexible and β is less than unity, they are assumed to consist of flexible (free) and rigid (complexed) sections. If the thermodynamic flexibility of the complexed sections is considerably lower than that of the free sections, and the interaction energy (ε) between units of the matrix is fairly high ($\geqq$ kT in absolute values), then an all-or-none-type complexation may be realized. This implies that there are two parts of matrices in the solution, one part being filled with oligomer almost completely ($\beta \simeq 1$) and another part being more or less free ($\beta \simeq 0$). The tendency for this type of complexation becomes stronger with rising ε and increasing difference in the flexibility between free and complexed sections or with decreasing ratio of the chain length of oligomer and matrix. These phenomena may be caused by long-range interactions (volume interactions), entropy (loss when interacting sections are arranged alongside one another), and other weak interactions such as stacking interactions and solvophobic interactions, and conformational changes of the matrix chain through complexation[505].

Kabanov and coworkers[183, 203] reported the distribution of oligomers on the matrix polymer in the system poly(carboxylic acid)s-PEO or PVPo. From the results of sedimentation diagrams (Figs. 43(a) and (b)), it is found that in aqueous solutions of mixtures of PMAA and PEO or PMAA and PVPo, the oligomers are distributed on the matrix chains according to the all-or-none-type principle whereas in the systems PMAA-PEO in water-methanol mixtures (methanol above 30 vol%), PAA-PEO and PAA-PVPo, the distribution of the oligomers is random. These results may be explained by the specific conformation (packed coil) of PMAA in water, i.e. when unstructured polymers (in this experiment PAA) are used as matrices, the oligomers are statisti-

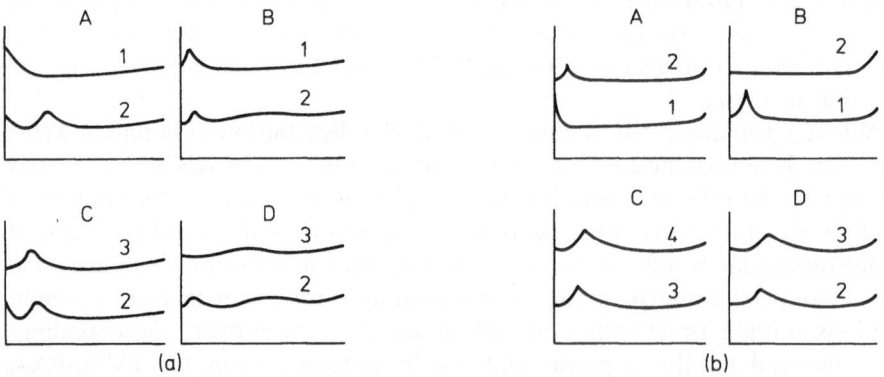

Fig. 43 a, b. Specificity of the formation of intermacromolecular complexes[203].
(a) Distribution of oligomers on matrix polymer chains in the "all or none" type. Sedimentation diagrams of the poly(methacrylic acid) (PMAA)-poly(ethylene oxide) (PEO) system in water. (1) [PEO]/[PMAA] = 0, (2) [PEO]/[PMAA] = 0.5, (3) [PEO]/[PMAA] = 1.0. C and D are continuations of A and B, respectively. (b) Random distribution of oligomers on matrix polymer chains. Sedimentation diagrams of the poly(acrylic acid) (PAA)-poly-(N-vinyl-2-pyrrolidone) (PVPo) system in water. (1) [PVPo]/[PAA] = 0, (2) [PVPo]/[PAA] = 0.25, (3) [PVPo]/[PAA] = 0.5, (4) [PVPo]/[PAA] = 1.0. B is a continuation of A

cally distributed on the matrix chains. Under these conditions, where the distribution according to the all-or-none-type principle would involve large entropy losses, the oligomer chains are distributed more or less uniformly. Due to complexation the matrix macromolecules are highly compact (see Chap. 3) and their intramolecular mobility decreases[506, 507].

4.2 Selective Intermacromolecular Complexation

In biological systems, a macromolecular chain effectively selects a complementary one to form an intermacromolecular complex. In this way, very specific functionalities become effective. Synthetic polymers can also form intermacromolecular complexes, but the ability of a synthetic polymer to select only one objective polymer as in biological systems has not yet been realized, except for several specific systems of pairs of polymers which include one of the complementary base pairs of nucleic acid individually, e.g. poly(A)-poly(U) and poly(I)-poly(C) (see Sect. 3.3). The intermacromolecular complex formation of synthetic polymers is controlled by many factors such as interaction forces, solvent, ionic strength, temperature, pH, etc. Moreover, the cooperative and concerted interactions of each active site play an important role in complex formation. These phenomena suggest that the selective intermacromolecular complexation can be realized under suitable conditions.

Selective interchain macromolecular complexation is described as follows:

$$P_1 + P_2 \longrightarrow (P_1 - P_2)_{complex} \tag{53}$$

$$P_3 + P_2 \longrightarrow (P_3 - P_2)_{complex} \tag{54}$$

$$P_1 + P_2 + P_3 \longrightarrow (P_1 - P_2)_{complex} + P_3 \tag{55}$$

P_1 and P_3 can interact with P_2 individually to form the intermacromolecular complex. However, when three polymer components coexist under suitable conditions, only P_1 preferentially forms a complex with P_2 in this system.

This type of selective complexation was discussed by Kabanov et al.[498]. When there exist P_1 and P_3 with the chain length ν_1 and ν_3, and the matrix P_2, the ratio of the degrees of filling of the matrix by P_1 and P_3, β_1/β_3, depends on the concentrations, chain lengths and chemical properties of P_1 and P_3. The ratio β_1/β_3 can be determined analytically in the case $\beta_{tot} = \beta_1 + \beta_3 \simeq 1$. The condition $\beta_{tot} \simeq 1$ is satisfied, when the chain lengths of P_1 and P_3 are sufficiently longer than the critical chain length and when the size of the matrix is inadequate to bind all the chains of both P_1 and P_3. Then,

$$\frac{\beta_1}{\beta_3} = \frac{C_1}{C_3}\left[\frac{K_1^{\nu_3}}{K_3^{\nu_1}}\right] = \psi\left[\frac{C_1}{C_3}\right] \tag{56}$$

where C_1 and C_3 are the equilibrium concentrations of free P_1 and P_3 in solution. The magnitude of the coefficient ψ characterizes the degree of selectivity, called "selective coefficient". If both the chain lengths of P_1 and P_3 are much longer than the critical chain length, since the effect of chain length on the interaction is minimized, then one can assume $\nu_1 \simeq \nu_3 \simeq \nu$,

$$\psi = \left[\frac{K_1}{K_3}\right]^\nu \tag{57}$$

$$K_1/K_3 = \exp(-\Delta\Delta F^0/RT) \tag{58}$$

$$\Delta\Delta F^0 = \Delta F_1^0 - \Delta F_3^0 \tag{59}$$

where ΔF_1^0 and ΔF_3^0 represent free energy changes of the complexation per mole of repeating unit for P_1–P_2 and P_3–P_2, respectively. When P_1 and P_3 have the same chemical structure, i.e. $K_1 = K_3 = K$ (different chain length), the selective coefficient is expressed as

$$\psi = K^{\Delta\nu} \tag{60}$$

Table 24. Selective complexation of macromolecules

Polymer			Solvent (pH)	Complex	Main controlling factor
P_1	P_2	P_3			
PHMPA	PMAA	PVPo	DMSO	PHMPA-PMAA	Solvation
P2VP	PMAA	PEO	H_2O (3)	PEO-PMAA	pH
			H_2O (6)	P2VP-PMAA	
PEI	PMAA	PEO	H_2O (2)	PEO-PMAA	pH
			H_2O (5)	PEI-PMAA	
			H_2O (9)	None	
PAAm	PMAA	PVA	H_2O (2)	PAAm-PMAA	
PAAm	PMAA	PEO	H_2O (2)	PAAm-PMAA	
PAAm	PMAA	PVPo	H_2O (2)	PAAm-PMAA	
PVPo	PMAA	PEO	H_2O (2)	PVPo-PMAA	Concerted effect
PVPo	PMAA	PVA	H_2O (2)	PVPo-PMAA	
PEO	PMAA	PVA	H_2O (2)	PEO-PMAA	
PMAA	PEO	PAA	H_2O (2)	PMAA-PEO	
PEO(I)	PMAA	PEO(II)	H_2O (2)	PEO(I)-PMAA	Cooperative effect
PMAA	PEO	PGA	H_2O (2)	PMAA-PEO	Steric effect
P4VP	PMAA	P2VP	H_2O/MeOH	P4VP-PMAA	

Molecular weight: PEO(I) ≫ PEO(II)

PHMPA = poly(N,N-dimethyl-N'N'N''N''-tetramethylenephosphoric triamide), PMAA = poly-(methacrylic acid), PVPo = poly(N-vinyl-2-pyrrolidone), P2VP = poly(2-vinylpyridine), PEO = poly(ethylene oxide), PEI = poly(ethylenimine), PAAm = poly(acrylamide), PVA = poly(vinyl alcohol), PAA = poly(acrylic acid), PGA = poly(L-glutamic acid), P4VP = poly(4-vinylpyridine)

where $\Delta\nu = \nu_1 - \nu_3$. Even if $\Delta\Delta F^0$ is small, e.g. in the region of a few calories, does the selectivity coefficient reach a value of 10^2–10^3 at $\Delta\nu \simeq 200$–300. On the other hand, when $\Delta\Delta F^0$ is relatively high, e.g. -100 cal/mol per repeating unit, the selectivity coefficient reaches $10^5 \sim 10^{10}$ even at degrees of polymerization of a few tens.

Table 24 shows typical examples of selective macromolecular complex formation[508]. In the system of poly(N,N,-dimethyl-N'N'N''N''-tetramethylenephosphoric triamide) (PHMPA), PMAA and PVPo, the PHMPA-PMAA complex is preferentially formed in DMSO because the proton-accepting power of PHMPA is larger than that of PVPo and the solvent molecules. In a system containing a weak polybase, e.g. poly(2-vinylpyridine) (P2VP) or poly-(ethylenimine) (PEI), a weak polyacid, e.g. PMAA, and a proton-accepting nonionic polymer, e.g. PEO, the selective macromolecular complex formation of different pairs of polymers is goverend by the pH of the solution.

(1) Acidic condition

$$\Big\{ \overset{|}{\underset{|}{O}} + \Big\{ COOH + \overset{|}{\overset{+}{NH_2}} \longrightarrow \overset{|}{\underset{|}{O}} \cdots HOOC\Big\} + \overset{|}{\overset{+}{NH_2}} \tag{61}$$

(2) Neutral condition

$$\overset{|}{\underset{|}{O}} + \Big\{ \overset{-COO^-}{\underset{-COOH}{}} + \overset{+NH_2}{\underset{NH}{}} \longrightarrow \overset{|}{\underset{|}{O}} + \Big\{ \overset{COO^- \cdots {}^+NH_2}{\underset{COO\cdots H \cdots NH}{}} \tag{62}$$

(3) Alkaline condition

$$\overset{|}{\underset{|}{O}} + \Big\{ COO^- + \overset{\backslash}{NH} \longrightarrow \text{No interaction} \tag{63}$$

(1) Under acidic conditions, weak polybases (P2VP or PEI) are almost all protonated but a weak polyacid (PMAA) is scarcely dissociated. Thus, the complex of PMAA with the proton-accepting polymers (PEO or PVPo), which may form hydrogen bonds with PMAA, is preferentially formed.
(2) At neutral pH, both a weak polyacid and a weak polybase are partially ionized, resulting in the formation of the polyelectrolyte complexes.
(3) At alkaline pH, a weak polyacid is almost completely dissociated while polybases are not protonated; thus, neither the polyelectrolyte complex nor the complex resulting from hydrogen bonds are formed.

Next, in the system of two proton-accepting polymers, e.g. poly(acryl-amide) (PAAm), poly(vinyl alcohol) (PVA), PVPo and PEO, and a proton-donating polymer, e.g. PMAA, selective macromolecular complex formation is realized as shown in Table 24. Thus, under these experimental conditions, the complexation abilities of these proton-accepting polymers with respect to PMAA (mainly due to hydrogen bonds) follows the order

$$PEO \ (\overline{M_w} = 1.4 \times 10^6) > PAAm > PVPo > PEO \ (\overline{M_w} = 2900) > PVA \qquad (64)$$

On the basis of these results, the complexation may be described by the following scheme:

$$\qquad (65)$$

PAAm and PVPo bind PMAA more strongly than PEO and PVA, since PAAm forms complexes with PMAA not only through hydrogen bonds but also through ion-dipole interactions between the partially protonated amide groups of PAAm and the C=O dipoles of the carboxy groups of PMAA. Moreover, also hydrophobic interactions are assumed to be involved in selective complexation. The hydrophobicity of PVPo is the largest of all these polymers. Thus, the interaction between PVPo and PMAA is larger than between other combinations. In the system of PMAA, PEO and PAA, the PMAA-PEO complex is formed preferentially because of the difference of the hydrophobic interactions between PMAA-PEO and PAA-PEO. On the other hand, PVA has a strong tendency to aggregate with itself. Therefore, the complex formation of PMAA with PVA is difficult because PVA aggregates must be destroyed.

The selective intermacromolecular complexation based on the difference of the chain length of PEO is also shown in Table 24. It has already been pointed out that a cooperative interaction between macromolecules is a very important factor in the formation of intermacromolecular complexes. This phenomenon is explained by Eq. (60).

The difference between α-helical structures and random coils plays an important role in the complexation occurring in the system PEO-PMAA-PGA. The distance between two carboxy groups, which are fixed on the α-helical structure of PGA, does not match with the distance between two ether oxygen atoms of PEO. Thus, PEO interacts preferentially with the carboxy groups of PMAA existing in a random structure.

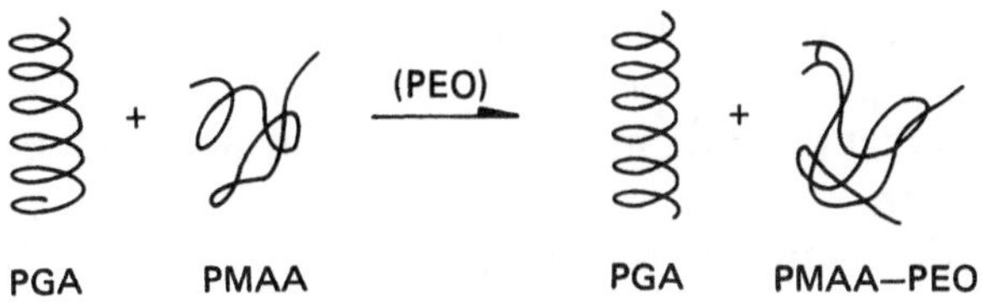

$$\text{(66)}$$

PGA   PMAA   PGA   PMAA—PEO

Steric hindrance also plays an important role in the complexation in the system of PMAA-P2VP-poly(4-vinylpyridine) (P4VP). Electrostatic interactions and hydrogen bonds appear to coexist in this system. The Coulomb force is here a comparatively long-range attraction force, the interaction force being inversely proportional to the square of the distance between two ionic sites. The hydrogen-bond interaction is a rather short-range interaction whereby steric factors are important. Then the position of nitrogen, which seems to be a binding center,in P2VP seems to make complexation difficult because it is so close to the main chain.

4.2.1 Macromolecular Substitution Reactions

As mentioned previously, selective intermacromolecular complexation is realized by the control of the difference of the total bond energy between each pair of polymers. Therefore, if P_1 can interact with P_2 more strongly than P_3 does, the interchain macromolecular substitution reaction of P_3 and P_1 takes place on the addition of P_1 to the P_2-P_3 complex solution. In these systems, it is expected that a cooperative interchain macromolecular substitution reaction of the type

$$(P_2\text{-}P_3)_{comp.} + P_1 \rightarrow (P_1\text{-}P_2)_{comp.} + P_3 \tag{67}$$

occurs. Such substitution reactions may proceed by the following mechanism:

$$(P_2\text{-}P_3)_{comp.} + P_1 \rightarrow (P_1\text{-}P_2\text{-}P_3)_{comp.} \rightarrow (P_1\text{-}P_2)_{comp.} + P_3 \tag{68}$$

This mechanism is illustrated by the following scheme:

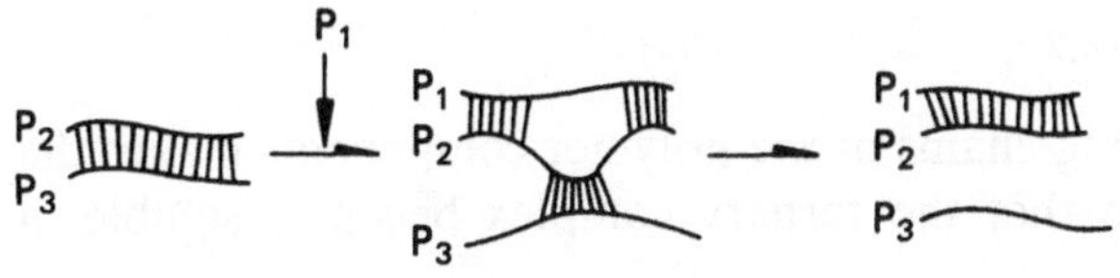

Before the complex once formed is completely destroyed, the third polymer component interacts with the first complex involving partial replacement to form a ternary polymer complex. Then, finally, the first polymer component is dissociated from the ternary complex. The substitution reaction, especially the release of the first polymer component, seems to be accelerated when the interaction forces between a parent polymer matrix and the polymer chain of the third component are stronger.

Kabanov et al.[188] studied the interchain substitution reaction

$$(PAA^*-PEO)_{comp.} + PMAA \rightarrow (PMAA-PEO)_{comp.} + PAA^* \tag{69}$$

by means of the polarized luminescence using PAA labelled with a lumines-cent probe (PAA*). When the reaction advances, $1/P^*$ increases from the value of PAA* in the complex to that of free PAA*. This reaction proceeds almost stoichiometrically. This result is confirmed by the theoretical conclu-sion drawn from Eqs. (57) and (60). Kabanov et al.[188] also reported the rate of the substitution reaction of PMAA with respect to PMAA* in the complex containing PEO (PMAA* is the labelled PMAA by the luminescent probe). When the chain length of PEO is slightly longer than the critical chain length, the reaction is completed immediately after mixing of the PMAA-PEO com-plex and PMAA*. The rate decreases with increasing chain length of PEO but finally, with much longer chain length of PEO the rate is independent of the chain length. These results may be explained by the assumption that, with short chain length, the substitution reaction takes place after almost all of the first complex (PMAA-PEO) is dissociated because of the easy dissociation of this complex. In contrast, PEO with longer chain length causes the substitution reaction to proceed via a ternary complex since dissociation of the first com-plex does practically not occur[509].

The results of the substitution reaction of the integral-type polycation (2 X)-PEO-PMAA system are shown in Fig. 44. The PEO-PMAA complex and the 2 X-PMAA complex are preferably formed in the low or high pH region, respectively, whereas in the intermediate pH region the 2 X-PEO-PMAA ternary complex is formed, but the yield of its precipitate is very low. Thus, in the 2 X-PEO-PMAA system, a substitution reaction takes place due to the process described by Eq. (68). The decrease in the yield of the precipi-tate of the ternary complex is well explained by the following scheme:

$$\tag{70}$$

PEO
PMAA ⟶ PEO
PMAA ⟶ PEO

              PMAA
2X ⟶ 2X ⟶ 2X

pH<3   3<pH<7   7<pH

Hence, part of the non-bonding chains of the polymer components makes the complex itself hydrophilic so that the ternary complex becomes soluble in aqueous medium.

---

* P = degree of polarization, $[P = (I_{\parallel} - I_{\perp})/(I_{\parallel} + I_{\perp})]$
$I_{\parallel}$ and $I_{\perp}$ are luminescence intenisties polarized parallel $(I_{\parallel})$ and perpendicular $(I_{\perp})$ to the polarization direction of the excitation light

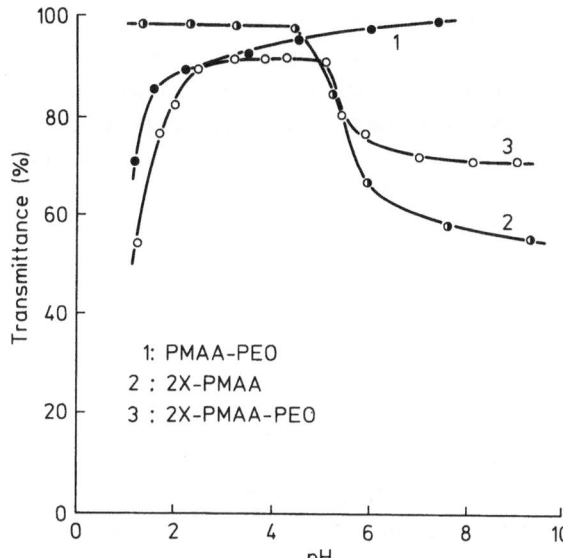

Fig. 44. Transmittance as a function of pH of substitution reactions in the integral-type polycation (2 X)-poly-(methycrylic acid) (PMAA)-poly-(ethylene oxide) (PEO) system

1: PMAA-PEO
2 : 2X-PMAA
3 : 2X-PMAA-PEO

As a result, a study of such selective intermacromolecular complexation and substitution reaction would be expected to clarify the fundamental phenomenon of the specific polymer reaction in biological systems and the polymer effects of synthetic macromolecules.

4.3 Supermolecular Structure of Intermacromolecular Complexes

In biological systems, complicated and specific functionalities are caused by the regular molecular aggregation, for example folding and renaturation and self-assembly of proteins. In recent years, extensive studies on the higher-order structures of biopolymers have been performed. However, a general rule for molecular aggregation has not yet been established even in biological systems because of the complexity and specificity of biopolymers. In contrast to biopolymers, synthetic polymers have rather simple structures. Thus, they are advantageously used for the detection and understanding of the fundamental phenomena of the complicated reactions occurring in vivo. For this reason, studies on the mechanisms of the complexations between synthetic macromolecules must give important suggestions for the design of functions of polymer chains. Such primary complexes obtained by the contact between synthetic polymer chain components undergo further aggregations under particular conditions like biological systems.

The macromolecular chains would form random coils or would become dispersed in the form of specific conformations. It seems that the formation of liquid crystalline layers of synthetic polypeptides is one of the examples of processes where ordered structures are spontaneously formed[510]. Aggregates could be dissolved only in good solvents whose enthalpy of mixing is negative or slightly positive. London interactions between polymers would prevent the particles from dispersing into the more stable solution by accelerating aggregation. It is expected that the complex has a random conformation in dilute solution. This means that polymer-polymer reactions proceed very rapidly so that the intermacromolecular complexes formed by the first contact of polymer chains with complementary polymer chains are considered to exist in a metastable state. The fact that the complex may exist in a non-equilibrium state may be explained as follows: since instantaneous complexation through secondary binding forces should give rise to distortions in conformations and weak interactions as well as main interactions, e.g. hydrophobic interactions, may act, so that the formed complexes would gradually transform into their more stable structures. If considerable conformational changes were permitted to occur easily, a regular helical-like conformation would result. In this case, if the interactions among macromolecules were not repulsive or at least very weak attractive forces, then precipitation resulting in phase separation would take place. If the concentration of the solute increases and the ellipticity of the complex aggregates is large, a tactoidal phase should appear and separate from the solution as a liquid crystalline state. Moreover, owing to the attractive forces between particles in a dilute nematic region, it would change into a fibrous aggregation state of polymers which is a highly and densely oriented state.

In such a way, an ordered supermolecular structure could be spontaneously formed from polymers exhibiting a random coil structure. The fundamental cause of this transformation is naturally the first-order structure of the polymer, i.e. its chemical structure. If a polymer with the proper structural units was prepared, then fiber formation would be accelerated and would especially account for the formation of a highly ordered structure.

4.3.1 Structure of Primary Complexes

Table 25[406] shows the results of X-ray diffraction analysis of polyelectrolyte components and their complexes[88]. PMAA shows only a halo ring at about $2\Theta = 20°$ which demonstrates that it is a completely amorphous polymer. In contrast, the polycation is a partially crystalline polymer, i.e. many diffraction rings are observed. Moreover, the following conclusions have been drawn:

(1) since the diffraction patterns are clearly recognizable, each crystalline part is comparatively perfect

(2) since many diffraction rings are observed, the symmetry of the crystals is low

(3) since the intensity of each diffraction ring is uniform, the orientation of the crystals is random.

Generally, in linear polymer systems, there are amorphous, semicrystalline and crystalline regions coexisting in a single polymer chain[511]. As shown in Table 25, the polyelectrolyte complex does not exhibit sufficient crystallinity

Table 25. Lattice spacing of polyelectrolyte complexes

Sample	Intensity	d(Å)	Sample	Intensity	d(Å)	Sample	Intensity	d(Å)
2X	w	2.21	Complex (I)	w	2.89		w	2.21
	s	2.89		m	3.67		s	2.68
	s	3.22		s	6.03	Complex (IV)	s	3.08
	m	3.61	Complex (II)	w	2.75		m	3.58
	m	4.08		m	3.76		w	4.26
	s	4.75		s	6.95		s	4.92
	w	6.03	Complex (III)	s	3.35			
	s	6.62		w	6.31			

Intensity: w = weak, m = medium, s = strong

Complex (I):   original complex obtained immediately after mixing of PMAA and 2X
Complex (II):   heating sample of complex (I)
Complex (III): complex in solution
Complex (IV): aggregated complex

but has a certain oriented structure different from that of the polymer components. Moreover, thermal treatment does not affect the crystallinity of the
polyelectrolyte complex. The crystallinity is assumed to be affected by the
following factors:
(1) the molar ratio of the polymer components
(2) the molecular weight and structure of the polymer components
(3) the crystallinity of the polymer components
(4) the intensity of interactions between the polymer components
(5) the conditions of the complex formation.

4.3.2 Association of Intermacromolecular Complexes

The precipitate of the polyelectrolyte complex formed by mixing aqueous
solutions of oppositely charged polyelectrolytes at concentrations higher than
10^{-1} (mol/l per repeating unit) assumes a poor crystalline structure. Such primary complexes linked by Coulomb forces are formed instantaneously and
irregularly. At lower concentration where precipitation of the complex is
avoided, the primary complexes exist as small particles in the solution. By
means of centrifugation, part of the complex is separated from the solution. If
the supernatant solution, which is clear, is kept under a nitrogen atmosphere at
room temperature, fine fibers are formed within about 10 days after mixing.
Figure 45 shows optical micrographs of the time-dependent aggregation of the
integral-type polycation (2X)-PMAA system. Within 4 days after mixing, the
fibrous aggregates are separated from the solution (a) and then they continue
to aggregate in all directions to a network structure (b). At last, these aggregates can be observed even with the naked eye (c). Many branches of the fine
fibers are clearly observed in Figs. 45(a) and (b). Such aggregation

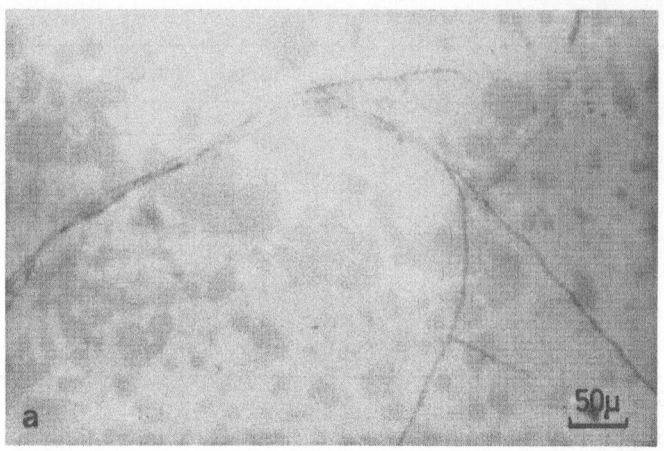

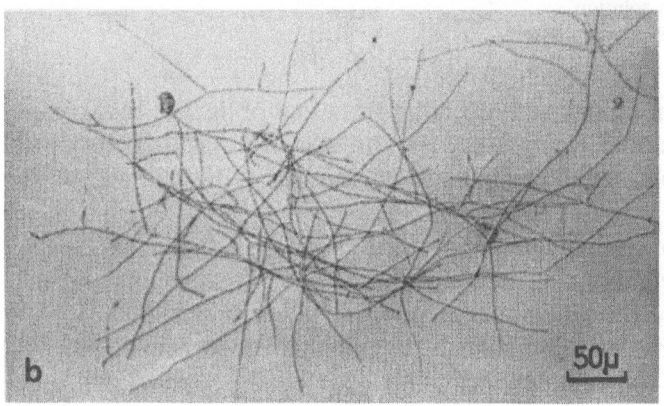

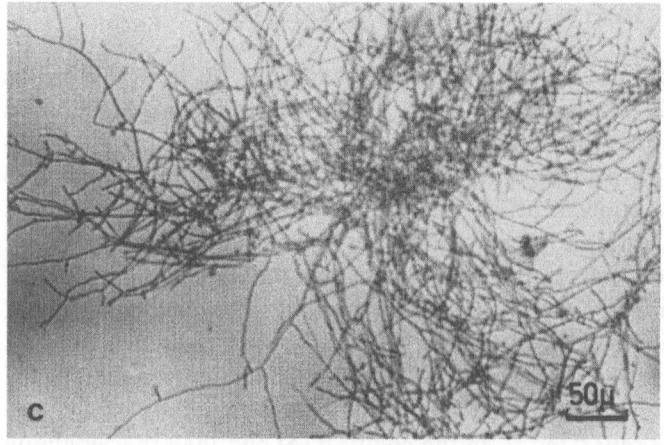

Fig. 45 a, b. Time dependence of the aggregation process of the polyelectrolyte complex composed of integral-type polycation (2 X)-poly(methacrylic acid) (PMAA) (× 125). (a) 4 days, (b) 7 days, (c) 14 days after mixing

phenomena are observed in the system of only one of the polyelectrolyte components but under conditions quite different from those of the complex systems. While the complex aggregates are obtained within about 4 days after mixing without microsalts, separation of the aggregates of polyelectrolytes in the presence of microsalts (0.2 mol/l) takes much longer time, i.e. more than a week. This salt effect coincides well with the dissociation states of the polyelectrolytes, i.e. aggregation of polyelectrolytes starts when the electrostatic repulsion is sufficiently weakened. Furthermore, the aggregation of the polyelectrolyte complex is accelerated by heating and stirring.

Firstly, conformational changes and/or the new formation of bonds in the complex (including rearrangements of bonds) were observed by means of circular dichroism and potentiometry as shown in Fig. 46. In the PGA-QOEI system, the α-helix of PGA is destabilized with aging. From the change in pH of the complex solution, it is found that even after the complex has been formed the further bindings are newly formed. In Fig. 46, ΔpH is the change of proton concentration calculated from the equation

$$\Delta pH = pH_i - pH_t \tag{71}$$

where pH_i and pH_t are the pH values measured before and 't' hours after mixing aqueous solutions of PAA and PVPo or an integral-type polycation (10,10-ionene), the initial pH of which are all denoted by pH_i. This means that in the PAA-PVPo system, the first contact between PAA and PVPo may occur at the surface of each polymer component domain and then complex formation is continued with time in the intersphere region of the primary complex taking up protons from the solution. In contrast, in the polyelectrolyte complex system, at the first contact of oppositely charged polyelectrolytes, complexation is

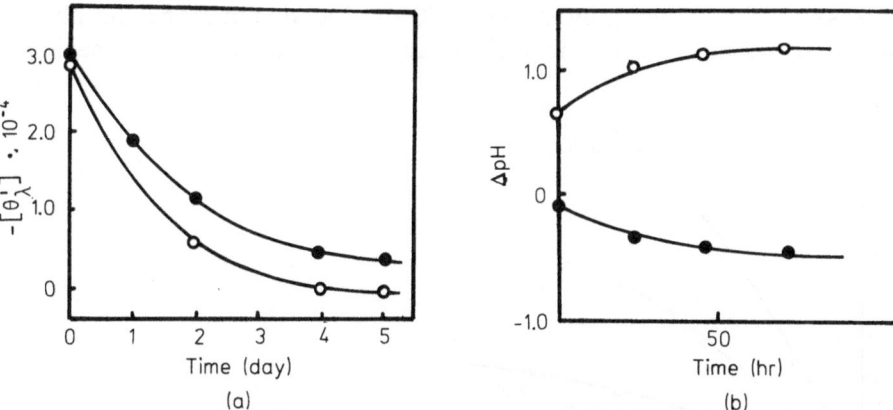

Fig. 46 a, b. Time dependence of the complexation. (a) conformational changes of the polymer components in the complexes ● Poly(L-glutamic acid) (PGA)-tetraethylenepentamine, ○ PGA-pentaethylenehexamine; (b) pH change of the complex solutions; ● Poly(acrylic acid) (PAA)-integral-type polycation (10,10-ionene) ○ PAA-poly(N-vinyl-2-pyrrolidone) (PVPo)

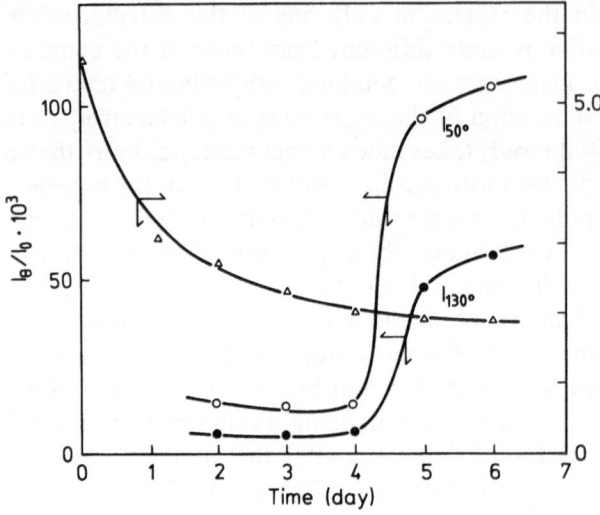

Fig. 47. Aggregation of a polyelectrolyte complex composed of poly(methacrylic acid) (PMAA) and an integral-type polycation (2 X) detected by means of the light scattering method

not completely finished and advances with time releasing HCl into the solution. Similar phenomena are observed in complexation systems of complementary polynucleotides[469].

Intercomplex aggregation can be also detected by the light scattering method. Figure 47 shows the time dependence of the changes of light scattering intensities at the angles of 50 and 130° and of their intensity ratios, $I_{50°}/I_{130°}$, of a polyelectrolyte complex composed of 2 X and PMAA. Light scattering intensity is proportional to the molecular weight and the volume of the polymer domain in solution, i.e. it increases with the rising molecular weight of the solute. The ratio $I_{50°}/I_{130°}$ means an asymmetric parameter (Z_{40}), i.e. if the solute has a spherical shape, it is equal to unity. When increasing the asymmetry of the solute, e.g. ellipsoid and rod, the value of this parameter increases. Light scattering intensities at 50 and 130° do not change for about 4 days after mixing, and then increase effectively to finally reach a maximum

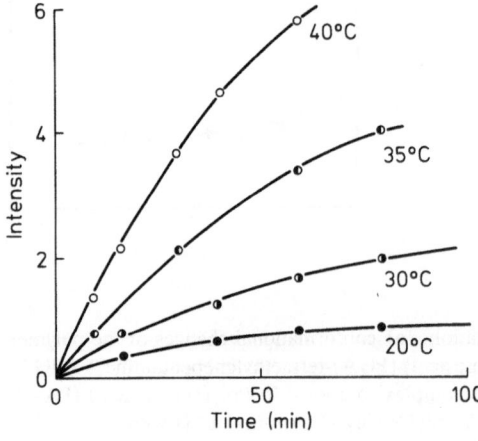

Fig. 48. Effect of temperature on the aggregation of the intermacromolecular complex of poly(methacrylic acid) (PMAA) and poly(ethylene oxide) (PEO) linked by hydrogen bonds

value. These results indicate that complex aggregation may start about 4 days after mixing, and are supported by the morphological findings mentioned previously (see Fig. 45). On the other hand, the asymmetry parameter decreases monotonously with time which means that the aggregate obtained after the first period exhibits a globular-like structure. Figure 48 shows the time dependence of the light scattering intensity in the PMAA-PEO system at various temperatures. Such intermacromolecular complexes formed by hydrogen bonds also aggregate with each other. At elevated temperatures, the aggregation of the complexes is accelerated. In comparison with polyelectrolyte complexes, the rate of aggregation of hydrogen-bonded complexes is much faster. From these results it is suggested that the driving force of the aggregation of the complex are partly hydrophobic interactions.

The formation process of polyelectrolyte complexes may be divided into three main classes (Fig. 49):
(1) primary complex formation
(2) reformation process within intracomplexes
(3) intercomplex aggregation process.

The first step is realized through secondary binding forces such as Coulomb forces immediately after mixing oppositely charged polyelectrolyte solutions. This reaction is very rapid. The second step proceeds within the order of an hour involving the formation of new bonds and/or the correction of the distortions of the polymer chains. The third step involves aggregation of secondary complexes mainly through hydrophobic interactions. Such aggregation is influenced by many factors, e.g. the structure of the polymer components and the complexation conditions.

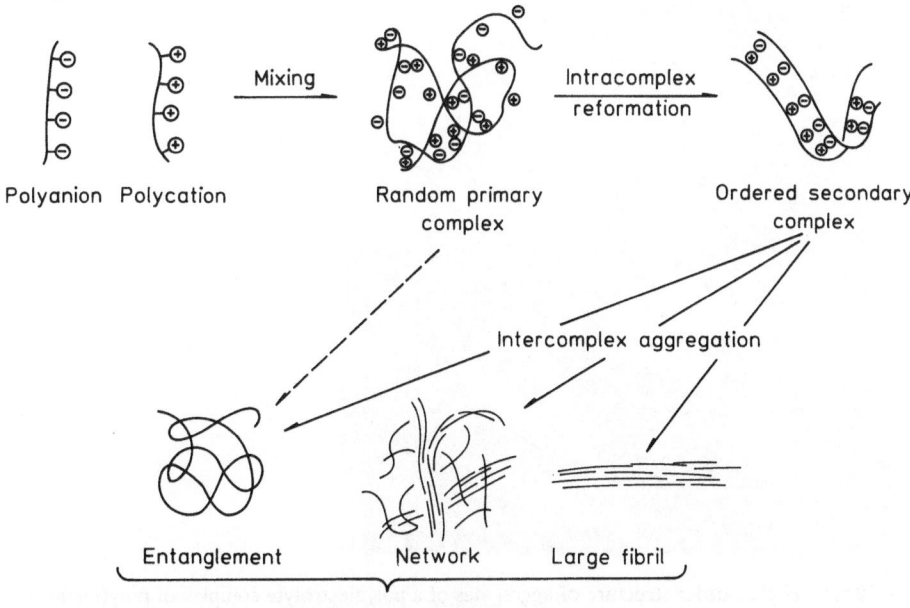

Fig. 49. Schematic representation of the aggregation of polyelectrolyte complexes

4.3.3 Supermolecular Structure of Aggregated Intermacromolecular Complexes

The final aggregates of the polyelectrolyte complexes are insoluble in ordinary solvents, and the molar ratio of the repeating units of the polymer components in the aggregates is almost unity. Furthermore, the final product is believed to be packed comparatively densely and regularly in contrast to the primary complex precipitates (see Table 25).

Several studies on the supermolecular structure of aggregated intermacromolecular complexes performed by optical and electron microscopies have been reported. The obtained results are summarized in Table 26. Under polarized light, assemblies of intermacromolecular complexes appear to be light, as shown in Fig. 50. From the results of X-ray analysis and polarized

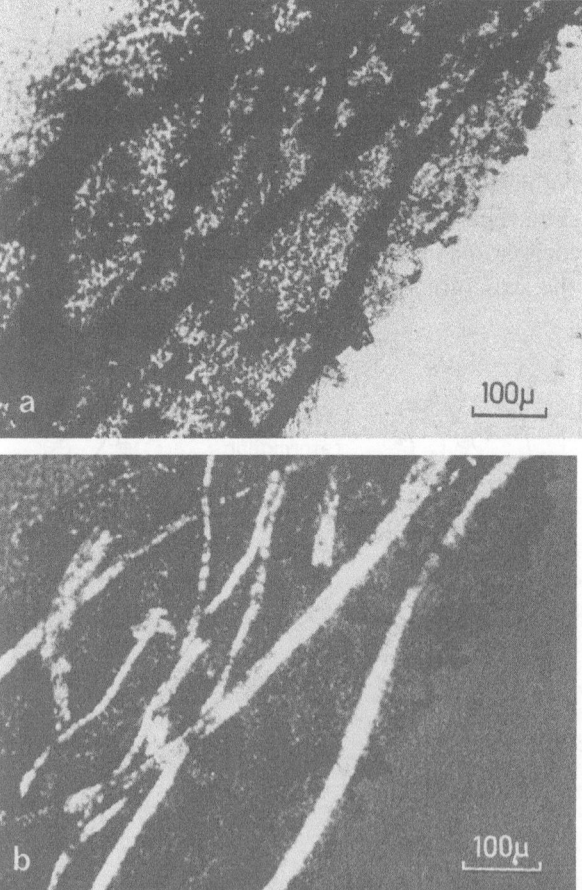

Fig. 50 a, b. Higher-order structure of aggregates of a polyelectrolyte complex of poly(methacrylic acid) (PMAA) and integral-type polycation (2X). (**a**) Optical micrograph, (**b**) polarized light micrograph

Table 26. Formation of higher-order aggregates of polymer complexes composed of synthetic polymers

Polymer complex	Morphology	Ref.
P4VP-DNA[a]	Fibrous (partially spherical)	512
PLL-PMAA	Curdy, fibrous	513
P4VP-PSS (Template polymerization)	Fibrous	514, 515
2X-PMAA	Fibrous (network)	128
$\left.\begin{matrix}2X\\3X\\6X\end{matrix}\right\}-\left\{\begin{matrix}PMAA\\PGA\end{matrix}\right.$	Fibrous (network)	126
PVBMA-PMAA	Needle-like	144
PLL-PAA	Globular	516, 517

[a] no synthetic deoxyribonucleic acid

P4VP = poly(4-vinylpyridine), PSS = poly(styrenesulfonic acid), PLL = poly(L-lysine), PMAA = poly(methacrylic acid), PGA = poly(L-glutamic acid), PVBMA = poly(4-vinylbenzyltrimethyl-ammonium chloride), 2X, 3X and 6X = integral-type polycations (see text)

light microscopy, it is suggested that the assembly forms a certain higher-order (supermolecular) structure. Moreover, it should be noted that solvent effects are also clearly observed. As stated previously, the polyelectrolyte complex prepared immediately after mixing the polyelectrolyte components in water is completely amorphous whereas the one obtained in methanol at 64.5 °C exhibits a fibrous structure with optical activity[518]. The supermolecular structure is, therefore, affected by the type of the polymer components used. The chemical environment can also control this structure. However, up to date, only morphological studies on the polyelectrolyte complex have been made. Thus, in the future, kinetic and thermodynamical aspects are also expected to be dealt with.

5 Future Development of Intermacromolecular Complexes Tailored to Special Uses

5.1 Intermacromolecular Interactions Applied to Ecological Protection

The formation of intermacromolecular complexes is considered to be very effective in removing and utilizing waste materials. In practice, polyelectrolytes are well-known as flocculants. Intermacromolecular complexes have the following advantages:

(1) Because the complexation reaction takes place almost quantitatively, it is suitable for the removal of very small quantities of organic materials.

(2) Since the product can be obtained as a precipitate under certain conditions, this method can readily be applied.

(3) Selection of appropriate polymer components allows a selective complex formation to be realized.

(4) Utilizing these phenomena skillfully, one can selectively recover a single organic material individually.

However, theoretical conclusions on the aggregation of waste materials with polyelectrolytes are insufficiently grounded in view of the fact that polyelectrolytes are widely used as flocculants[519]. This method may be applied to protein fractionation which is of fundamental importance in preventing medicine and clinical treatment[593, 594].

Moreover, intermacromolecular complex formation is applied to selective recovery of organic and metallic ions. For example, as shown in Table 27, Cu^{2+} ion is much more effectively precipitated by the polyelectrolyte complex than by one of its components[520]. Furthermore, polyelectrolyte complexes including some metal ions have been studied in recent years (see Sect. 3.2.). Crown ethers can bind certain cations; they especially exhibit high affinity to K^+. Smid et al.[521] synthesized poly(vinylbenzo-[18]-crown-6). Such polymers containing crown ether with K^+ behave like polycations in solution and can interact with polyanions such as poly(carboxylic acid) to generate a kind of polyelectrolyte complexes. Moreover, PAA may interact with the ether oxy-

Table 27. Adsorption of Cu^{2+} ion on a polyelectrolyte complex

Polymer	Yield (%)	Adsorption (%)
Poly(sodium styrenesulfonate)	0	0
Partially quaternized poly(4-vinylpyridine)	5	11
Polyelectrolyte complex[a]	84.3	91.7

[a] The polyelectrolyte complex is prepared by mixing poly(sodium styrenesulfonate) (NaSS) and partially quaternized poly(4-vinylpyridine) (QPVP) at [NaSS]/[cationic site of QPVP] = 1.0

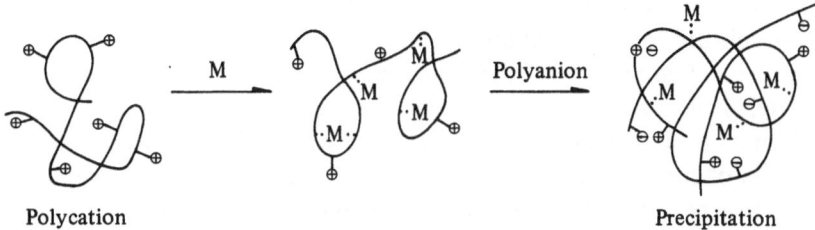

Polycation   Precipitation

gen of poly(vinylbenzo-[18]-crown-6) through hydrogen bonds. Kabanov et al.[522-525] described three-component polyelectrolyte complexes containing poly(carboxylic acid), poly(4-vinylpyridine) and Cu^{2+}. The effect of pH on two types of the metal-containing polyelectrolyte complexes was discussed (i.e. the change of the dissociation state of polyelectrolyte components) on the basis of the results of ESR measurements. Such intermacromolecular complexes with coordinated metal ions are expected to decisively affect the development of new polymer catalysts, redox systems and membranes with selective permeability to organic and inorganic materials as well as selective adsorption resins, because the doamin of the intermacromolecular complex provides a specific reaction field, e.g. spatially fixed ligands, hydrophobic doamin and so on.

5.2 Ultrafiltration Membranes

It is expected that the intermacromolecular complexes display entirely new physical and chemical characteristics different from those of the individual polymer components. So the following applications are, for example, considered: membranes for dialysis, ultrafiltration, fuel cells and battery separators, wearing apparel, electrically conductive and antistatic coatings for textiles, medical and surgical prosthetic materials, environmental sensors or chemical detectors, and electrodes modified with specific polymers.

As noted previously (see Chap. 3), the membranes of intermacromolecular complexes (especially polyelectrolyte complexes) have specific characteristics such as permeability to water, small molecules and gases. Therefore, ultrafiltration membranes are most useful applications of these intermacromolecular complexes. Considerable interest has consequently developed in the use of these membranes in hemodialyzers (artificial kidneys) and hemooxygenators (artificial lungs) whose efficiency is limited by already existing membrane materials which provide the major resistance to mass transfer.

5.3 Utilization of Special Local Fields in the Domains of Intermacromolecular Complexes

Intermacromolecular complexes are expected to control the properties of the domains through the following points
(1) the special local environment (for example, hydrophobic, electrostatic and conductive fields) is easily controlled in the domain of the complex
(2) several kinds of active sites may be oriented within the domain
(3) some kinds of functional groups are easily introduced into the complex.

Actually, coacervate complexes have been focused as prebiological systmes[526], since the coacervate complex with a certain interface around the domain shows the functionalities, e.g. the formation of a stable local environment, adsorption and the concentration of compounds, specific transport processes and the chemical potential. Some enzyme reactions occurring in such coacervate complexes were studied as a model system of chemical evolution by Oparine et al., for example, the preparation and decomposition of starch[527, 528], the synthesis of poly(adenilic acid)[529, 530], and redox reactions[531, 532]. Moreover, coacervate complexes are recently focused as artificial cells.

5.4 Mechanochemistry

Mechanochemical reactions (conversion of mechanical to chemical energy and vise versa) have widely been studied. They are caused by proton transfer[153, 533], ion exchange[534, 535], chelate formation[270, 536], redox reaction [537, 538], phase transitions[539, 540], cis-trans transitions[541], and formation of intermacromolecular complexes[185].

Osada[202] described the mechanochemical system of poly(methacrylic acid)-poly(ethylene oxide) by utilizing intermacromolecular complexation between the two component polymers (Fig. 51). The PMAA membrane is contracted by more than 90% of its original length when heated from 10 to 60 °C. When decreasing the temperature, the membrane almost completely recovers its original state. The work spent per a contraction is about 5×10^{-3} cal/g of membrane.

Mechanochemistry is very important and interesting in the development of a "clean energy". However, up to date, there are some problems awaiting solution, e.g. the energy taken from the mechanochemical engine is small, the

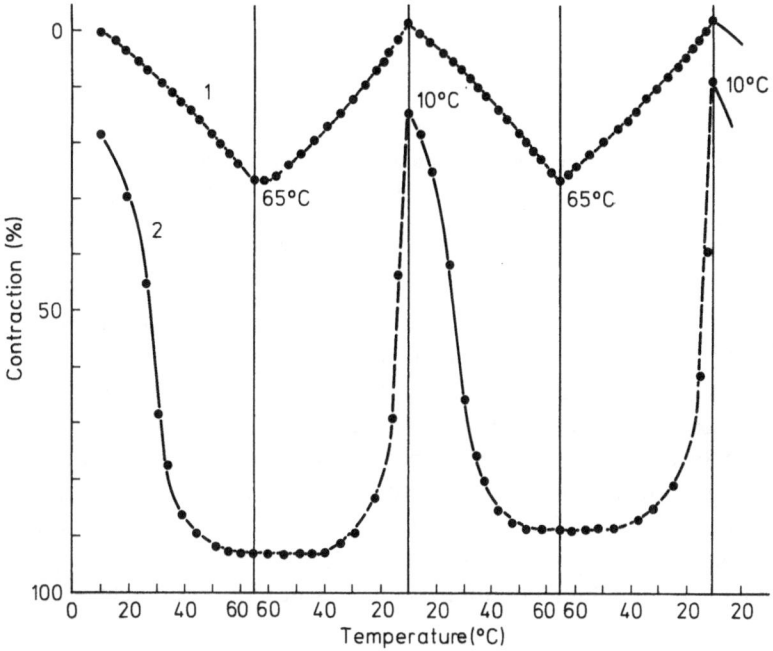

Fig. 51. Contraction of a poly(methacrylic acid) (PMAA) membrane in the presence of poly-(ethylene oxide) (PEO) in the steeping solution as a function of temperature[202]. (*1*) In the absence of PEO, (*2*) In the presence of PEO (0.015 mol of repeating unit/l)

response time is long, expansion and contraction are not completely reversible.

5.5 Matrix Polymerization

Polymerization of monomers in the presence of polymers which can interact with monomers or newly formed polymers via secondary binding forces is called matrix polymerization (or template polymerization, replica polymerization) (Fig. 52). It is expected that matrix polymerization fundamentally affects the kinetic behavior and/or controls some structural details (for example, molecular weight and its distribution, tacticities, optical isomerism, etc.). However, a variation of structural details seems to be realized only when the geometry of the monomers firmly bound on the matrix polymers, which exhibit a regular arrangement of their structures, can be controlled.

Up to now, many works on matrix polymerization have been reported, for example, monomers and matrix polymers can interact with each other through electrostatic ineraction[542, 543], charge-transfer interaction [544, 545], hydrogen bonding[232, 546], van der Waals force[547, 548], and covalent bonds[549]. Blumstein et al. analyzed the effect of microsalts and solvent on the rate of polymerization of ionizable monomers in the presence of polyelectrolytes[550-552]. Challa et al.

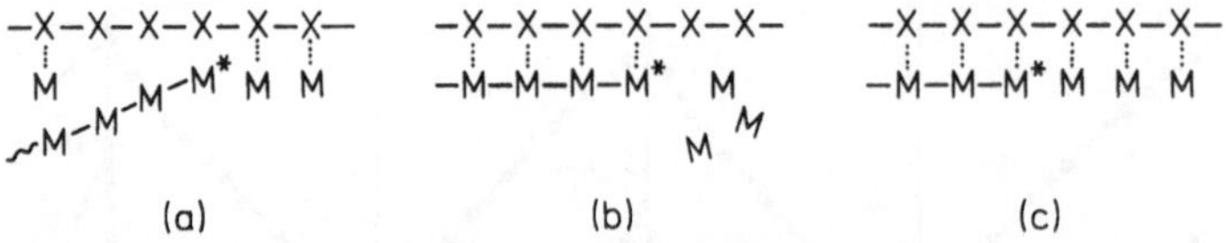

```
 -X-X-X-X-X-X-         -X-X-X-X-X-X-         -X-X-X-X-X-X-
 ⋮   ⋮⋮  ⋮ ⋮           ⋮  ⋮ ⋮⋮    ⋮           ⋮ ⋮ ⋮  ⋮ ⋮ ⋮
 M   M-M-M* M M       -M-M-M-M-M* M         -M-M-M* M M M
  ⌣M-M-M-M*                        M
                                    M M
                                       M
        (a)                    (b)                    (c)
```

Fig. 52 a–c. Schematic representation of three types of interactions between matrix polymer and monomer or growing chain in matrix polymerization. (**a**) Monomers are adsorbed preferentially on the matrix polymer chain, (**b**) Growing chains are preferentially adsorbed on the matrix polymer chain, (**c**) Monomers and growing chains are both adsorbed on the matrix polymer chain

reported the stereoregularity of the daughter polymer in stereocomplex system[553]. The matrix effect of water on the polymerization of acrylonitrile[554], acrylic acid[555] and on the polycondensation of polyamides[556, 557] has also been discussed. Moreover, matrix effect of synthetic polymers containing nucleic acid bases on the polycondensation of nucleotides or polyamides[558–560] are being studied. It is interesting that poly(acrylic acid) polymerized on chitosan reveals optical activity and stereoregularity[561]. However, successful control of the structural regulation or transcription of the information of parent polymer chains to daughter polymer chains has not yet been achieved.

5.6 Applications of Macromolecular Complexes and Molecular Assemblies

A large number of tests has been carried out by Bixler[54] and Refojo[56] to evaluate energetically the general biocompatibility of polyelectrolyte complexes as well as their specific compatibility with blood components. Extract and implant tests revealed that polyelectrolyte complexes display no toxicity[562, 563]. Values of hemoglobin released from erythrocytes are low, indicating that hemolysis is insignificant. Sakurai et al.[81–83] studied quantitatively the amount and denaturation of serum proteins adsorbed on polyelectrolyte complexes and concluded that moderately anionic complexes show the best compatibility with blood components[564]. Table 28 shows the results of *in vivo* clotting tests for polyelectrolyte complexes, silicon rubber, teflon and polypropylene by Bixler et al.[65]. It was concluded that the moderately anionic polyelectrolyte complexes possess highly significant non-thrombogenic activity, superior to that display by conventional low-surface energy polymers. Kikuchi et al.[565–568] reported the relation between the significant thromboresistance of polysaccharide-polysaccharide complexes and their structures, they pointed out the importance of the polysaccharide species used and the method of the complex preparation. Since heparinized materials display good blood compatibility, many studies have been performed on the preparation of the heparin-bound surface due to Coulomb forces on hydrogels with quaternary ammonium residues[569–571].

Due to their limited mechanical strength, polyelectrolyte complexes are usually used as coating agents on fabrics and other supports. Thus, it is neces-

Table 28. In vivo clotting test of the polyelectrolyte complex of poly(sodium styrenesulfonate)-poly(4-vinylbenzyl-trimethylammonium chloride) in comparison with other prosthetic rings[65]

Ring Type	Ionic Structure	H_2O Content	2-Hour Implant Results	2-Week Implant Results
Polyelectrolyte complex ring				
I. Moderately anionic	0.5 meq. excess polyanion per dry gram of resin	55% (wet basis)	○□ ○□ ○□	○□ ○□ ○□ ○□ ○□
II. Highly anionic	1.3 meq. anionic excess	80%	●□ ●□ ●■	
III. neutral	neutral	50%	○□ ○□ ●□	
IV. Moderately cationic	0.86 meq. cationic excess	67%	●■ ●■ ●■	
Prosthetic ring				
Silicone rubber	—		●■ ●■ ●■	
Teflon	—		●■ ●■ ●■	
Polypropylene	—		●■ ●■ ●■	

Side view   *End view*

□ atrial end   ○ atrial end

sary to prepare a suitable substrate for the coating with polyelectrolyte complexes[572]. Other difficulties have also been found to arise in the sterilization of polyelectrolyte complexes. Autoclave sterilization of these complexes can result in disintegration of the gel structure. In gas sterilization entrapped ethylene oxide may be left in the complex. However, an advantage of these materials is that the net charge of the system (anionic or cationic) can easily be controlled. This is achieved by adding stoichiometrically larger amounts of one of the two polymeric components to the equimolar complex during complex formation[573].

The interactions of synthetic macromolecules with cell membranes provide many information which are useful for a great number of applications; for example cell protectors cell dispersive supporters, fusogen, bactericides, cell separators, cell-cultured material and so on. In recent years, interferon has been recognized as an antivirally active and cancer-depressing material. Interferon is induced *in vivo* by foreign substances, such as viruses, some low molecular weight compounds, and synthetic macromolecules. The high molecular weight polyriboinosinic-polyribocytidylic acid double stranded helix is now known as one of the excellent synthetic interferon inducers[574, 575]. Recently, Papahadjopoulos et al.[576] have demonstrated the enhancement of interferon induction *in vitro* by using poly(I)-poly(C)-entrapped phospholipid vesicles. Moreover, other synthetic macromolecules such as polyanions (e.g. poly(carboxylic acid)s and polysaccharides) and weak polybases (e.g. polyethyleneimine and basic homopolypeptides) are also considered to be useful.

Nowadays cell technology as well as gene manipulation are being watched with keen interest. Cell fusion has been directly applied to the creation of new intermediate hybrid cells, especially of higher plants. Many excellent reviews of the cell fusion of higher plants have been published[577, 578]. Cell fusion is also important in the injection of several substances, e.g. drugs, proteins, nucleic acids, etc. into culture cells (microinjection). This cell fusion is induced by viruses, low molecular weight salts, amphiphiles, watersoluble macromolecules and so on. Specific macromolecules such as poly(ethylene oxide) (PEO) and dextran give rise to the fusion of some cell lines causing slight cytoplasmic lysis. Since PEO is relatively nontoxic, it markedly promotes chemically induced cell fusion and has a wide range of applications: PEO may be highly valuable in obtaining hybrid cells for use in a wide range of biochemical and genetic investigations. Though the mechanism of PEO-induced cell fusion has not yet been clarified, this technique provides increasing applications in many fields. To develop it further, intermacromolecular interactions must be studied more extensively, because essentially the interactions between macromolecules and molecular assemblies (cell membranes and tissue) are important.

Other applications of intermacromolecular interaction and complexes are picked up as follows. Some water-soluble nonionic polymers have been applied to the isolation of some serum proteins from human serum under mild conditions different from former techniques such as Cohn method[579] and methods using inorganic salts[580]. Poly(ethylene oxide) with high molecular weight has a higher degree of selectivity for specific proteins[581–583]. On the other hand, polymers containing 7,7,8,8-tetracyanoquinodimethane (TCNQ) exhibit electrical conductivity[584]. This means that the conductivity of TCNQ fixed on polycations through electrostatic interaction is controlled by the structure of polycations. It should be noted that macromolecules not only promote molecular ordering but also impair molecular packing. Polyelectrolyte complexes are increasingly used as commercial products such as membranes for dialysis[434], ultrafiltration[585] and reverse osmosis[586], dry hydrogels[587], blood-compatible materials[588], light sensitive catalysis[589], and dry-charged automotive batteries[590, 591].

In spite of the importance for these characteristic and interesting applications, new developments as well as fundamental and systematic studies of this field are still insufficient. The authors believe that active works in this field should be taken and epoch-making results must establish novel field in polymer science in the near future.

6 References

1. Stockmayer, W. H.: Makromol. Chem. *35*, 54 (1960)
2. Yamakawa, H.: J. Chem. Phys. *48*, 2103 (1968)
3. Vollmert, B., Stutz, H.: Angew. Makromol. Chem. *20*, 71 (1971)
4. Watson, J. D., Crick, F. H. C.: Nature *171*, 738 (1953)
5. Coiro, V. M. et al.: J. Polym. Sci., *C16*, 4591 (1969)
6. Flory, P. J.: J. Am. Chem. Soc. *87*, 1838 (1965)
7. Prigonine, I.: The Molecular Theory of Solutions, Interpress, New York 1957
8. Patterson, P., Delmas, G.: Trans. Far. Soc. *65*, 708 (1969)
9. Kolnibolotchuk, N. K., Klenin, V. J., Frenkel, S. Ya.: J. Polym. Sci., Symp. *44*, 119 (1974)
10. Klenin, V. J. et al.: J. Polym. Sci., Symp. *44*, 131 (1974)
11. Colwell, C. E., Livengood, S. M.: J. Soc. Cosm. Chem. *13*, 201 (1962)
12. Barone, G. et al.: J. Polym. Sci., Symp.*44*, 1 (1974)
13. Yuan, L., Veis, A.: Biopolymers *12*, 1437 (1973)
14. Imai, N.: Suppl. Prog. Theory Phys. *17*, 54 (1961)
15. Meeten, G. H.: Polymer *15*, 187 (1974)
16. Heitz, F., Marchal, E., Spach, G.: Macromolecules *8*, 145 (1975)
17. Prins, W.: Macromolecules *7*, 527 (1974)
18. Kwei, T. K., Nishi, T., Roberts, R. F.: Macromolecules *7*, 667 (1974)
19. Stockmayer, W. H.: J. Phys. Chem. *9*, 398 (1941)
20. Lippincott, E. R.: J. Chem. Phys. *23*, 1099 (1955)
21. Scheraga, H. A.: Biochemistry *6*, 3791 (1967)
22. London, F.: Z. Phys. *63*, 245 (1930)
23. Mulliken, R. S.: J. Am. Chem. Soc. *74*, 811 (1952)
24. Kauzmann, W.: Adv. Protein Chem. *14*, 1 (1959)
25. Frank, H. S., Evans, M. W.: J. Chem. Phys. *13*, 507 (1945)
26. Nemethy, G., Scheraga, H. A.: J. Chem. Phys. *36*, 3401 (1962)
27. Tanford, C.: J. Am. Chem. Soc. *84*, 4240 (1962)
28. Siegel, B., Breslow, R.: J. Am. Chem. Soc. *97*, 6869 (1975)
29. Holins, K. G. et al.: Nature *254*, 192 (1975)
30. Casper, D.: Adv. Protein Chem. *18*, 37 (1963)
31. Champness, J. N. et al.: Nature *259*, 20 (1976)
32. Durham, A. C., Finch, J. S., Klug, A.: Nature, New Biology *229*, 37 (1971)
33. Flory, P. J.: Science *124*, 53 (1956)
34. Oosawa, F., Asakura, S., Ooi, T.: Suppl. Prog. Theory Phys. *17*, 14 (1961)
35. Oosawa, F., Maruyama, M., Fujime, S.: J. Theor. Biol. *36*, 203 (1972)
36. Asakura, S.: J. Mol. Biol. *35*, 237 (1968)
37. Fischer, E.: Chem. Ber. *27*, 2985 (1894)
38. Koschland, D. E., Jr.: Enzymes (2nd edit.) *1*, 305 (1959)
39. Lumry, R.: Enzymes (2nd edit.) *1*, 157 (1959)
40. Singer, S. J., Nicolson, G. L.: Science *175*, 720 (1972)

41. Utsumi, H., Tunggal, B. D., Stoffel, W.: Biochemistry *19*, 2385 (1980)
42. Stollery, J. G. et al.: Biochemistry *19*, 2391 (1980)
43. Kang, S. Y., Gutowsky, H. S., Oldfield, E.: Biochemistry *18*, 3286 (1979)
44. Cheng, C., Chan, S. I.: Biochemistry *13*, 4381 (1974)
45. Hammes, G. G., Schullery, S. E.: Biochemistry *9*, 2555 (1970)
46. Yu, K. Y., Baldassare, J. J.: Biochemistry *13*, 4375 (1974)
47. Jost, P. C. et al.: Proc. Nat. Acad. Sci. USA *70*, 480 (1973)
48. Rembaum, A.: Appl. Polym. Symp. *22*, 299 (1973)
49. Michaels, A. S., Miekka, R. G.: J. Phys. Chem. *65*, 1765 (1961)
50. Michaels, A. S., Mir, L., Schneider, N. S.: J. Phys. Chem. *69*, 1447 (1965)
51. Michaels, A. S., Falkenstein, G. L., Schneider, N. S.: J. Phys. Chem. *69*, 1456 (1965)
52. Michaels, A. S.: Ind. Eng. Chem. *57*, 32 (1965)
53. Bixler, H. J., Michaels, A. S.: Encycl. Polym. Sci. Technol. *10*, 765 (1969)
54. Vogel, M. K. et al.: J. Macromol. Sci., Chem. *A 4*, 675 (1970)
55. Refojo, M. F.: J. Appl. Polym. Sci. *11*, 1991 (1967)
56. Refojo, M. F.: J. Appl. Polym. Sci. *9*, 3417 (1965)
57. Reid, S. H. et al.: J. Colloid Interface Sci. *26*, 222 (1968)
58. Michaeli, I., Bejerano, T.: J. Polym. Sci., Part C *22*, 909 (1969)
59. Yasuda, H., Ikenberry, L. D., Lammaze, C. E.: Makromol. Chem. *125*, 108 (1969)
60. Hoffman, A., Lewis, R., Michaels, A. S.: Accounts Polym. Prep. *10*, 916 (1969)
61. Duby, J., Prijot, E.: Arch. Opht. (Paris) *29*, 393 (1969)
62. Refojo, M. F. et al.: Arch. Opht. (Chicago) *80*, 645 (1968)
63. Mishima, S.: Arch. Opht. (Chicago) *73*, 233 (1965)
64. Blatt, W. F. et al.: Science *150*, 3693 (1965)
65. Markley, I. L., Bixler, H. J., Cross, R. A.: J. Biomed. Mat. Res. *2*, 145 (1968)
66. Saraves, C. A. et al.: Science *150*, 224 (1965)
67. Yodh, S. B., Wright, R. L.: J. Neurosurgery *26*, 504 (1967)
68. Wakoe, J. R. H., Posner, A. M.: Nature *213*, 692 (1967)
69. Koshikawa, S.: Kobunshi *21*, 512 (1972)
70. Smolen, V. F., Hahman, D. E.: J. Colloid Interface Sci. *42*, 70 (1973)
71. Nelson, L. et al.: Surgery *67*, 826 (1970)
72. Bruck, S. D.: J. Biomed. Mat. Res. *7*, 387 (1973)
73. Yano, O., Wada, Y.: J. Appl. Polym. Sci. *25*, 1723 (1980)
74. Abe, M. et al.: Denki Kagaku *48*, 412 (1980)
75. Kurokawa, Y. et al.: J. Appl. Polym. Sci. *25*, 1645 (1980)
76. Kataoka, K. et al.: J. Biomed. Mat. Res. *14*, 817 (1980)
77. Kataoka, K. et al.: Makromol. Chem. *181*, 1363 (1980)
78. Nakajima, A., Sato, H.: Bull. Inst. Chem. Res. Kyoto Univ. *47*, 177 (1969)
79. Miwa, M., Sanada, K., Tsuchida, E.: Nippon Kagaku Kaishi *1972*, 2161 (1972)
80. Fuoss, R. M., Sadek, H.: Science *110*, 552 (1949)
81. Kataoka, K. et al.: Makromol. Chem. *179*, 1121 (1978)
82. Akaike, T. et al.: ACS Polym. Prep. *20*, 585 (1979)
83. Kataoka, K. et al.: ACS Polym. Prep. *20*, 581 (1979)
84. Yano, O., Wada, Y.: Rep. Prog. Polym. Phys. Japan *21*, 303 (1978)
85. Yano, O., Wada, Y.: Rep. Prog. Polym. Phys. Japan *22*, 349 (1979)
86. Tsuchida, E., Osada, Y., Sanada, K.: J. Polym. Sci., Part A-1 *10*, 3397 (1972)
87. Shirakawa, N. et al.: Technol. Rep., Tohoku Univ. *44*, 503 (1979)
88. Kurokawa, Y. et al.: Technol. Rep., Tohoku Univ. *49*, 129 (1979)
89. Zezin, A. B. et al.: Vysokomol. Soyed. *A 14*, 772 (1972)
90. Valuyeva, S. P., Zezin, A. B., Savin, V. A.: Vysokomol. Soyed. *A 16*, 212 (1974)
91. Kuznetosova, N. L. et al.: Vysokomol. Soyed. *A 16*, 2435 (1974)
92. Factor, A., Rouse, T. O.: J. Electrochem. Soc. *127*, 1313 (1980)
93. Factor, A., Heinsohn, G. E.: Polym. Lett. *9*, 289 (1971)
94. Akahoshi, H., Toshima, S., Itaya, K.: J. Phys. Chem. *85*, 818 (1981)

95. Worsfold, D. J.: J. Polym. Sci., Polym. Chem. Ed. *12*, 337 (1974)
96. Kakivaya, S. R., Blumstein, A.: J. Chem. Sci., Chem. Commun. *1974*, 459 (1974)
97. Yui, N., Kurokawa, Y., Miura, E.: Denki Kagaku *41*, 613 (1973)
98. Saito, M. et al.: Nippon Kagaku Kaishi *1974*, 977 (1974)
99. Abe, K., Tsuchida, E.: Makromol. Chem. *176*, 803 (1975)
100. Suzuki, Y., Kurokawa, Y., Yui, N.: Technol. Rep., Tohoku Univ. *37*, 173 (1972)
101. Matsumoto, T. et al.: Nippon Kagaku Kaishi *1974*, 178 (1974)
102. Sato, H., Nakajima, A.: Polym. J. *9*, 241 (1975)
103. Rogacheva, V. B., Zezin, A. B.: Vysokomol. Soyed. *B 11*, 327 (1969)
104. Rogacheva, V. B., Zezin, A. B., Kargin, V. A.: Vysokomol. Soyed. *B 12*, 826 (1970)
105. Miura, E., Kurokawa, Y., Yui, N.: Technol. Rep., Tohoku Univ. *38*, 213 (1973)
106. Zezin, A. B. et al.: Vysokomol. Soyed. *A 17*, 2637 (1975)
107. Philipp, B. et al.: Acta Polym. *30*, 563 (1979)
108. Nakanishi, Y. et al.: Technol. Rep., Tohoku Univ. *44*, 137 (1979)
109. Polderman, A.: Biopolymers *14*, 2181 (1975)
110. Philipp, B. et al.: Acta Polym. *31*, 654 (1980)
111. Shirakawa, N. et al.: Technol. Rep. Tohoku Univ. *45*, 71 (1980)
112. Nakanishi, Y., Kurokawa, Y., Yui, N.: Denki Kagaku *48*, 449 (1980)
113. Kurokawa, Y., Shirakawa, N., Yui, N.: Kobunshi Ronbunshu *37*, 503 (1980)
114. Philipp, B. et al.: Eur. Polym. J. *17*, 615 (1981)
115. Kalyuzhnaya, R. I. et al.: Vysokomol. Soyed. *A 18*, 71 (1976)
116. Djadoun, S., Goldberg, R. N., Morawetz, H.: Macromolecules *10*, 1015 (1977)
117. Nakajima, T., Hattori, H.: Rep. Prog. Polym. Phys. Japan *21*, 193 (1978)
118. Nakajima, T. et al.: Rep. Prog. Polym. Phys., Japan *22*, 157 (1979)
119. Lutsenko, V. V., Zezin, A. B., Rudman, A. R.: Vysokomol. Soyed. *B 13*, 396 (1971)
120. Plate, N. A., Alieva, E. D., Kalachev, A. A.: Vysokomol. Soyed. *A 23*, 640 (1981)
121. Osada, Y., Abe, K., Tsuchida, E.: Nippon Kagaku Kaishi *1973*, 2219 (1973)
122. Osada, Y., Abe, K., Tsuchida, E.: Nippon Kagaku Kaishi *1973*, 2222 (1973)
123. Tsuchida, E., Osada, Y., Abe, K.: Makromol. Chem. *175*, 583 (1974)
124. Tsuchida, E., Osada, Y.: Makromol. Chem. *175*, 593 (1974)
125. Tsuchida, E.: Makromol. Chem. *175*, 603 (1974)
126. Tsuchida, E., Abe, K., Honma, M.: Macromolecules *9*, 112 (1976)
127. Abe, K., Ohno, H., Tsuchida, E.: Makromol. Chem. *178*, 2285 (1977)
128. Tsuchida, E., Osada, Y.: Kobunshi Kagaku *30*, 517 (1973)
129. Rogacheva, V. B., Mirlina, S. Ya., Kargin, V. A.: Vysokomol. Soyed. *B 12*, 340 (1970)
130. Gluyaeva, Zh. G. et al.: Vysokomol. Soyed. *A 16*, 1852 (1974)
131. Kabanov, V. A. et al.: Dokl. Akad. Nauk SSSR *230*, 139 (1976)
132. Savinova, I. V. et al.: Vysokomol. Soyed. *A 18*, 2050 (1976)
133. Gluyaeva, Zh. G. et al.: Vysokomol. Soyed. *A 18*, 2800 (1976)
134. Furuta, T., Kurokawa, Y., Yui, N.: Technol. Rep., Tohoku Univ. *38*, 207 (1973)
135. Yui, N., Kurokawa, Y., Iwabuchi, C.: Denki Kagaku *43*, 71 (1975)
136. Yui, N., Kurokawa, Y.: Bunseki Kiki *12*, 492 (1974)
137. Tsuchida, E., Osada, Y.: Kobunshi *22*, 384 (1973)
138. Philipp, B. et al.: Acta Polym. *31*, 592 (1980)
139. Chepurov, A. K. et al.: Polym. Med. *10*, 243 (1980)
140. Legkanets, R. E., Izotov, M. Z., Bekturov, E. A.: Izv. Akad. Nauk Kaz. SSR, Ser. Khim. *1981(3)*, 42 (1981)
141. Shayakhmetov, Sh. Sh. et al.: Izv. Akad. Nauk Kaz. SSR, Ser. Khim. *1981(3)*, 47 (1981)
142. Iwabuchi, C., Kurokawa, Y., Yui, N.: Technol. Rep., Tohoku Univ. *38*, 493 (1973)
143. Tsuchida, E., Osada, Y., Abe, K.: J. Polym. Sci., Polym. Chem. Ed. *14*, 767 (1976)
144. Salamone, J. C. et al.: Polymer *20*, 611 (1979)
145. Izumrudov, V. A. et al.: Vysokomol. Soyed. *A 20*, 400 (1978)
146. Starodubtsev, S. G., Kabanov, V. A.: Vysokomol. Soyed. *A 19*, 1948 (1977)
147. Zezin, A. B., Lutsenko, V. V., Kabanov, V. A.: Vysokomol. Soyed. *A 16*, 600 (1974)

148. Izumrudov, V. A. et al.: Vysokomol. Soyed. *A 22*, 692 (1980)
149. Margolin, A. L. et al.: Biochim. Biophys. Acta *660*, 359 (1981)
150. Izumrudov, V. A., Zezin, A. B.: Vysokomol. Soyed. *A 18*, 2488 (1976)
151. Kalyuzhnaya, R. I. et al.: Vysokomol. Soyed. *A 17*, 2786 (1975)
152. Matsumoto, T.: Kobunshi Kagaku *13*, 132 (1956)
153. Okihana, H., Nakajima, A.: Bull. Inst. Chem. Res., Kyoto Univ. *54*, 63 (1976)
154. Nakajima, A., Sato, H.: Biopolymers *11*, 1345 (1972)
155. Sato, H., Nakajima, A.: Colloid Polym. Sci. *252*, 944 (1974)
156. Sato, H., Maeda, M., Nakajima, A.: J. Appl. Polym. Sci. *23*, 1759 (1979)
157. Sato, H., Nakajima, A.: Colloid Polym. Sci. *252*, 294 (1974)
158. Nakajima, A., Sato, H.: Bull. Inst. Chem. Res., Kyoto Univ. *52*, 664 (1974)
159. Hosono, M. et al.: Rep. Poval Commitee, Kyoto *61*, 79 (1972)
160. Hosono, M., Kusudo, O., Tsuji, W.: Rep. Poval Commitee, Kyoto *57*, 99 (1970)
161. Hosono, M. et al.: Rep. Poval Commitee, Kyoto *63*, 61 (1974)
162. Hosono, M. et al.: Bull. Inst. Chem. Res., Kyoto Univ. *52*, 442 (1974)
163. Hosono, M. et al.: Kobunshi Ronbunshu *33*, 509 (1976)
164. Sakurada, I., Hosono, M., Tamamura, S.: Bull. Inst. Chem. Res., Kyoto Univ. *42*, 145 (1964)
165. Hosono, M. et al.: Bull. Inst. Chem. Res., Kyoto Univ. *51*, 104 (1973)
166. Kokufuta, E.: Macromolecules *12*, 350 (1979)
167. Terayama, H.: J. Polym. Sci. *8*, 243 (1952)
168. Terayama, H.: Nippon Kagaku Zasshi *78*, 1261 (1957)
169. Kokufuta, E., Shimizu, H., Nakamura, I.: Macromolecules *14*, 1178 (1981)
170. Terayama, H.: J. Polym. Sci. *20*, 477 (1956)
171. Kharenko, A. V. et al.: Vysokomol. Soyed. *A 18*, 1604 (1976)
172. Reinert, K. E. W.: Bioelectrochem. Bioenerg. *8*, 301 (1981)
173. Kabanov, V. A. et al.: Makromol. Chem., Rapid Commun. *2*, 343 (1981)
174. Philipp, B. et al.: Z. Anorg. Allg. Chem. *479*, 219 (1981)
175. Bailey, Jr., F. E., France, H. G.: J. Polym. Sci. *49*, 397 (1961)
176. Bailey, Jr., F. E., Ludberg, R. D., Callard, R. W.: J. Polym. Sci., Part-A *2*, 845 (1964)
177. Osada, Y.: J. Polym. Sci., Polym. Lett. Ed. *18*, 281 (1980)
178. Smith, K. L., Winslow, A. E., Peterson, D. E.: Ind. Eng. Chem. *51*, 1361 (1959)
179. Papisov, I. M. et al.: Vysokomol. Soyed. *A 15*, 2003 (1973)
180. Bailey, Jr., F. E., Ludberg, R. D., Callard, R. W.: ACS Polym. Prep. *1*, 202 (1960)
181. Osada, Y. et al.: Dokl. Akad. Nauk. SSSR *191*, 3996 (1970)
182. Antipina, A. D., Papisov, I. M., Kargin, V. A.: Vysokomol. Soyed. *B 12*, 329 (1970)
183. Papisov, I. M. et al.: Dokl. Akad. Nauk SSSR *199*, 1364 (1971)
184. Antipina, A. D. et al.: Vysokomol. Soyed. *A 14*, 941 (1972)
185. Saito, H., Osada, Y.: Nippon Kagaku Kaishi *1976*, 171 (1976)
186. Osada, Y., Sato, M.: Nippon Kagaku Kaishi *1976*, 175 (1976)
187. Papisov, I. M. et al.: Vysokomol. Soyed. *A 16*, 1133 (1974)
188. Papisov, I. M. et al.: Dokl. Akad. Nauk SSSR *214*, 861 (1974)
189. Ikawa, T. et al.: J. Polym. Sci., Polym. Chem. Ed. *13*, 1505 (1975)
190. Osada, Y., Sato, M.: Makromol. Chem. *176*, 2761 (1975)
191. Osada, Y., Sato, M.: J. Polym. Sci., Polym. Lett. Ed. *14*, 129 (1976)
192. Osada, Y. et al.: Vysokomol. Soyed. *A 14*, 2462 (1973)
193. Papisov, I. M. et al: Dokl. Akad. Nauk SSSR *208*, 397 (1973)
194. Bekturov, E. A., Bimendina, L. A., Saltybaeva, S. S.: Makromol. Chem. *180*, 1813 (1979)
195. Bimendina, L. A. et al.: Izv. Akad. Nauk, Kaz. SSR, Ser. Khim. *26(4)*, 57 (1976)
196. Ferguson, J., McLeod, C.: Europ. Polym. J. *7*, 71 (1966)
197. Bimendina, L. A., Saltybaeva, S. S., Bekturov, E. A.: Izv. Akad. Nauk, Kaz. SSR, Ser. Khim. *28(6)*, 22 (1978)
198. Kim, H. J., Tonami, H.: Kobunshi Ronbunshu *35*, 395 (1978)
199. Papisov, I. M., Baranovskii, V. Yu., Kabanov, V. A.: Vysokomol. Soyed. *A 17*, 2428 (1975)

200. Osada, Y., Saito, H.: Nippon Kagaku Kaishi *1976*, 832 (1976)
201. Osada, Y.: J. Polym. Sci., Polym. Chem. Ed. *17*, 3485 (1979)
202. Osada, Y.: J. Polym. Sci., Polym. Chem. Ed. *15*, 255 (1977)
203. Papisov, I. M., Baranovskii, V. Yu., Kabanov, V. A.: Vysokomol. Soyed. *A 17*, 2104 (1975)
204. Baeras, G. et al.: Dokl. Akad. Nauk SSSR *241*, 95 (1978)
205. Frolova, V. A. et al.: Izv. Akad. Nauk, Kaz. SSR, Ser. Khim. *28(4)*, 43 (1978)
206. Litmanovich, A. A.: Vestn. Mosk. Univ., Ser. 2, Khim. *19*, 617 (1978)
207. Miyauchi, S., Ikeda, K., Tanzawa, H.: Rep. Eng., Niigata Univ. *26*, 105 (1977)
208. Subramanian, R., Natarajan, P.: J. Polym. Sci., Polym. Chem. Ed. *17*, 1855 (1979)
209. Wilson, R. W., Bloomfield, V. A.: Biopolymers *18*, 1205 (1979)
210. Tamaki, K. et al.: Kobunshi Ronbunshu *35*, 525 (1978)
211. Osada, Y., Sato, M.: Polymers *21*, 1057 (1980)
212. Litmanovich, A. A., Kazarin, L. A., Papisov, I. M.: Vysokomol. Soyed. *B 18*, 681 (1976)
213. Zheltonozhskaya, T. B., Eremenko, B. V., Uskov, I. A.: Dopov. Akad. Nauk Ukr. RSR, Ser. B: Geol., Khim. Biol. Nauki *1980(12)*, 43 (1980)
214. Zheltonozhskaya, T. B.: Deposite Doc., VINITI 3784, 426 (1979)
215. Tanaka, T. et al.: J. Macromol. Sci., Phys. *B 17*, 723 (1980)
216. Tsutsui, T., Tanaka, R.: Kobunshi Ronbunshu *37*, 603 (1980)
217. Ohno, H., Matsuda, H., Tsuchida, E.: Makromol. Chem. *182*, 2267 (1981)
218. Osada, Y., Takeuchi, Y.: J. Polym. Sci., Polym. Lett. Ed. *19*, 303 (1981)
219. Baranovskii, V. Y. et al.: Vysokomol. Soyed. *B 22*, 854 (1980)
220. Saltybaeva, S. S., Bimendina, L. A., Bekturov, E. A.: Izv. Akad. Nauk Kaz. SSR, Ser. Khim. *1981(1)*, 24 (1981)
221. Lipatov, Yu. S. et al.: Dopov. Akad. Nauk Ukr. RSR, Ser. B: Geol., Khim. Biol. Nauki. *1981(2)*, 52 (1981)
222. Ohno, H., Abe, K., Tsuchida, E.: Makromol. Chem. *179*, 755 (1978)
223. Bimendina, L. A., Roganov, V. V., Bekturov, E. A.: J. Polym. Sci., Symp. *44*, 65 (1974)
224. Tsutsui, T. et al.: Kobunshi Ronbunshu *35*, 517 (1978)
225. Bimendina, L. A., Roganov, V. V., Bekturov, E. A.: Vysokomol. Soyed. *A 16*, 2810 (1974)
226. Ohno, H., Nii, A., Tsuchida, E.: Makromol. Chem. *181*, 1227 (1980)
227. Ohno, H., Tsuchida, E.: Makromol. Chem., Rapid Commun. *1*, 591 (1980)
228. Dobry, A., Kawenoki, F. B.: Bull. Soc. Chim. Berg. *57*, 280 (1948)
229. Neel, J., Sebille, B.: Compt. Rend. Acad. Sci *250*, 1052 (1960)
230. Nagai, K., Nabekura, K.: Kobunshi Ronbunshu *37*, 73 (1980)
231. Kawenoki, F. B.: Compt. Rend. Acad. Sci. *262*, 278 (1966)
232. Ferguson, J., Shah, S. A.: Europ. Polym. J. *4*, 343 (1968)
233. Bartels, T., Tan, Y. Y., Challa, G.: J. Polym. Sci., Polym. Chem. Ed. *15*, 341 (1977)
234. Lorenz, D. H.: Encycl. Polym. Sci. Technol. *14*, 243 (1971)
235. Sakaguchi, Y. et al.: Kobunshi Kagaku *27*, 284 (1970)
236. Kirsh, Yu. E., Soos, T. A., Karaputadze, T. M.: Europ. Polym. J. *15*, 223 (1979)
237. Ferguson, J., McLeod, C.: Europ. Polym. J. *10*, 1083 (1973)
238. Molyneux, B. P., Frank, H. P.: J. Am. Chem. Soc. *83*, 3169 (1961)
239. Morishima, Y., Fujisawa, K., Nozakura, S.: J. Polym. Sci., Polym. Lett. Ed. *14*, 467 (1976)
240. Abe, K. et al.: Makromol. Chem. *179*, 2043 (1978)
241. Kawenoki, F. B.: Compt. Rend. Acad. Sci., Ser. C, *263*, 203 (1966)
242. Sistel, C. L., Sebille, B.: J. Polym. Sci., Symp. *52*, 311 (1975)
243. Bimendina, L. A., Bekturov, E. A., Roganov, V. V.: Chem. Zvesti *30*, 301 (1976)
244. Ferguson, J., Rajan, V. S.: Europ. Polym. J. *15*, 627 (1979)
245. Bimendina, L. A., Tleubaeva, G. S., Bekturov, E. A.: Vysokomol. Soyed. *A 19*, 71 (1977)
246. Bimendina, L. A. et al.: J. Polym. Sci., Symp. *66*, 9 (1979)
247. Tleubaeva, G. S., Bimendina, L. A., Bekturov, E. A.: Izv. Akad. Nauk, Kaz. SSR, Ser. Khim. *28(2)*, 42 (1978)
248. Bekturov, E. A., Bimendina, L. A.: Izv. Akad. Nauk, Kaz. SSR, Ser. Khim. *28(1)*, 68 (1978)

249. Tleubaeva, G. S. et al.: Izv. Akad. Nauk, Kaz. SSR, Ser. Khim. *28(3)*, 66 (1978)
250. Neel, J., Sebille, B.: Compt. Rend. Acad. Sci. *250*, 1270 (1960)
251. Litmanovich, A. A., Papisov, I. M., Kabanov, V. A.: Vysokomol. Soyed. *A 22*, 1180 (1980)
252. Frolova, V. A., Bimendina, L. A., Bekturov E. A.: Izv. Akad. Nauk Kaz. SSR, Ser. Khim. *1979(5)*, 32 (1979)
253. Litmanovich, A. A. et al.: Vysokomol. Soyed. *B 22*, 236 (1980)
254. Bimendina, L. A. et al.: Izv. Akad. Nauk Kaz. SSR, Ser. Khim. *1980(4)*, 86 (1980)
255. Slatybaeva, S. S., Bimendina, L. A., Bekturov, E. A.: Izv. Akad. Nauk Kaz. SSR, Ser. Khim. *1980(4)*, 37 (1980)
256. Saltybaeva, S. S., Bimendina, L. A., Bekturov, E. A.: Izv. Akad. Nauk Kaz. SSR, Ser. Khim. *1980(3)*, 35 (1980)
257. Tleubaeva, G. S., Bimendina, L. A., Bekturov, E. A.: Izv. Akad. Nauk Kaz. SSR, Ser. Khim. *1980(6)*, 69 (1980)
258. Subramanian, R., Natarajan, P.: Makromol. Chem., Rapid Commun. *1*, 47 (1980)
259. Higashi, F., Taguchi, Y.: J. Polym. Sci., Polym. Chem. Ed. *18*, 2875 (1980)
260. Alberda van Ekenstein, G. O. R., Koetsier, D. E., Tan, Y. Y.: Europ. Polym. J. *17*, 845 (1981)
261. Tleubaeva, G. S. et al.: Izv. Akad. Nauk Kaz. SSR, Ser. Khim. *1981(1)*, 28 (1981)
262. Williamson, G. K., Wright, B.: J. Polym. Sci., Part A *3*, 3885 (1965)
263. Okhrimenko, I. S., Dyakonova, E. B.: Vysokomol. Soyed. *6*, 1891 (1964)
264. Dyakonova, E. B., Okhrimenko, I. S., Yefremov, I. F.: Vysokomol. Soyed. *7*, 1016 (1965)
265. Okhrimenko, I. S. et al.: Vysokomol. Soyed. *8*, 1707 (1966)
266. Distler, G. I. et al.: Vysokomol. Soyed. *8*, 1737 (1966)
267. Frank, A.: Makromol. Chem. *96*, 258 (1966)
268. Kuhn, W.: Makromol. Chem. *35*, 200 (1960)
269. Kuhn, W., Thurkauf, M.: Kolloid Z-Z Polym. *184*, 114 (1963)
270. Kuhn, W., Toth, I., Kuhn, H. J.: Makromol. Chem. *60*, 77 (1963)
271. Kuhn, W. et al.: Nature *165*, 514 (1950)
272. Kawakami, H., Kawashima, K.: Kobunshi Kagaku *17*, 316 (1960)
273. Kawakami, H., Kawashima, K.: Kobunshi Kagaku *17*, 273 (1960)
274. Kawakami, H., Kawashima, K.: Kobunshi Kagaku *17*, 319 (1960)
275. Kawakami, H., Mori, N., Kawashima, K.: Kobunshi Kagaku *17*, 485 (1960)
276. Horiuchi, H.: Kobunshi Ronbunshu *36*, 287 (1979)
277. Horiuchi, H.: Kobunshi Ronbunshu *34*, 521 (1977)
278. Fedotov, N. G., Vedeneev, V. I., Sarkisov, O. M.: Dokl. Akad. Nauk SSSR *208*, 401 (1973)
279. Horiuchi, H., Morisawa, K.: Kobunshi Ronbunshu *37*, 9 (1980)
280. Horiuchi, H.: Kobunshi Ronbunshu *37*, 543 (1980)
281. Murohashi, S., Kuwabara, A.: Nagaoka Kogyo Koto Senmon Gakko Kenkyu Kiyo *17*, 17 (1981)
282. Horiuchi, H., Ohshita, T.: Kobunshi Ronbunshu *38*, 407 (1981)
283. Gandurin, L. I., Smirnova, E. A., Bulysheva, L. K.: Izv. Vyssh. Uchebn. Zaved., Tekhnol. Tekst. Prom-sti *1979(6)*, 68 (1979)
284. Nikolaev, A. F. et al.: Vysokomol. Soyed. *B 21*, 723 (1979)
285. Klenina, O. V., Fain, E. G.: Vysokomol. Soyed. *A 23*, 1298 (1981)
286. Otocka, E. P., Eirich, F. R.: J. Polym. Sci., Part A-2 *6*, 895 (1968)
287. Otocka, E. P., Eirich, F. R.: J. Polym. Sci., Part A-2 *6*, 913 (1968)
288. Anufrieva, E. et al.: Makromol. Chem. *180*, 1843 (1979)
289. Anufrieva, E. et al.: Vysokomol. Soyed. *A 14*, 1430 (1972)
290. Anufrieva, E. et al.: Dokl. Akad. Nauk SSSR *220*, 353 (1975)
291. Ohno, H., Tsuchida, E.: Polym. Prep. Japan *28*, 1055 (1979)
292. Ohno, H., Nii, A., Tsuchida, E.: Polym. Prep. Japan *27*, 964 (1978)
293. Chatterjee, S. K., Prokopova, E., Bohdanecky, M.: J. Macromol. Sci., Phys. *B 16*, 9 (1979)
294. Bimendina, L. A., Frolova, V. A., Bekturov, E. A.: Izv. Akad. Nauk, Kaz. SSR, Ser. Khim. *27*, 66 (1977)

295. Jaycox, G. D., Smid, J.: Makromol. Chem., Rapid Commun. *2*, 499 (1981)
296. Matsubayashi, K., Hirano, Y.: Seni Gakkaishi *17*, 637 (1960)
297. Staszewska, D., Bohdanecky, M.: Europ. Polym. J. *17*, 245 (1981)
298. Godovskii, Yu. K. et al.: Vysokomol. Soyed. *A 21*, 127 (1979)
299. Vorenkamp, E. J., Bosscher, F., Challa, G.: Polymer *20*, 59 (1979)
300. Challa, G., De Boer, A., Tan, Y. Y. : Internatl. J. Polym. Mat. *4*, 239 (1976)
301. Liquori, A. M. et al.: Nature *206*, 358 (1965)
302. Kusanagi, H., Tadokoro, H., Chatani, Y.: Macromolecules *9*, 531 (1976)
303. Buter, R., Tan, Y. Y., Challa, G.: J. Polym. Sci., Part A-1 *10*, 1031 (1972)
304. Buter, R., Tan, Y. Y., Challa, G.: J. Polym. Sci., Polym. Chem. Ed. *11*, 1003 (1973)
305. Buter, R., Tan, Y. Y., Challa, G.: J. Polym. Sci., Polym. Chem. Ed. *11*, 1013 (1973)
306. Spévačèk, J., Schneider, B.: Makromol. Chem. *176*, 729 (1975)
307. Feitsma, E. L., De Boer, A., Challa, G.: Polymer *16*, 515 (1975)
308. De Boer, A., Challa, G.: Polymer *17*, 633 (1976)
309. Pyrlik, M., Rehage, G.: Colloid Polym. Sci. *254*, 329 (1976)
310. Pyrlik, M., Rehage, G.: Rheor. Acta *14*, 303 (1975)
311. Mori, Y., Tanzawa, H.: J. Appl. Polym. Sci. *20*, 1775 (1976)
312. Mori, Y.: Dr. Thesis, Waseda University 1978
313. Mekenitskaya, L. I., Golova, L. K., Amerik, Yu. B.: Vysokomol. Soyed. *A 18*, 1799 (1976)
314. Kusakov, M. M., Mekenitskaya, L. I.: Vysokomol. Soyed. *B 15*, 213 (1973)
315. Sakai, Y., Tanzawa, H.: J. Appl. Polym. Sci. *22*, 1805 (1978)
316. Fujishige, S., Geoldi, P., Elias, H. G.: J. Macromol. Sci., Chem. *A 5*, 1011 (1971)
317. Fox, T. G. et al.: J. Am. Chem. Soc. *80*, 1768 (1958)
318. Schulz, G. V., Wunderlich, W., Kirste, R.: Makromol. Chem. *75*, 22 (1964)
319. Watanabe, W. H. et al.: J. Phys. Chem. *65*, 896 (1961)
320. Ryan, Ch. F., Fleisher, Jr., P. C.: J. Phys. Chem. *69*, 3384 (1965)
321. van den Berg, W. B. et al.: Nature *217*, 949 (1968)
322. Borchard, W., Pyrlik, M., Rehage, G.: Makromol. Chem. *145*, 169 (1971)
323. Chiang, R. et al.: J. Phys. Chem. *70*, 3591 (1966)
324. Liquori, A. M. et al.: J. Polym. Sci., Part A-2 *6*, 509 (1968)
325. Liquori, A. M., De Santis, M., D'Alagni, M.: J. Polym. Sci., Part B *4*, 943 (1966)
326. Beredjik, A. et al.: J. Polym. Sci. *46*, 268 (1960)
327. Liquori, A. M., De Santis Savino, M., D'Alagni, M.: Polym. Lett. *4*, 943 (1966)
328. Mihailov, M. et al.: Macromolecules *6*, 511 (1973)
329. Strupe, J. D., Hughes, R. E.: J. Am. Chem. Soc. *80*, 2341 (1958)
330. Schneider, B. et al.: Macromolecules *4*, 715 (1971)
331. Buter, R., Tan, Y. Y., Challa, G.: Polymer *14*, 171 (1973)
332. Borchard, W. et al.: Angew. Makromol. Chem. *29/30*, 471 (1973)
333. Buter, R., Tan, Y. Y., Challa, G.: J. Polym. Sci., Polym. Chem. Ed. *11*, 2975 (1973)
334. Biros, J., Masa, Z., Pouchly, J.: Europ. Polym. J. *10*, 629 (1974)
335. Pyrlik, M. et al.: Angew. Makromol. Chem. *36*, 133 (1974)
336. Dayatis, J., Reiss, C., Benoit, H.: Makromol. Chem. *120*, 113 (1968)
337. Gons, J., Vorenkamp, J. E., Challa, G.: J. Polym. Sci., Polym. Chem. Ed. *13*, 1699 (1975)
338. Spévàček, J., Schneider, B.: J. Polym. Sci., Polym. Lett. Ed. *12*, 349 (1974)
339. Liu, H. Z., Liu, K. J.: Macromolecules *1*, 157 (1968)
340. Spévàček, J., Schneider, B.: Makromol. Chem. *175*, 2939 (1974)
341. Tadokoro, H. et al.: Macromolecules *3*, 441 (1970)
342. Tadokoro, H. et al.: J. Polym. Sci. Polymer Chem. Ed. *11*, 825 (1973)
343. Grignor'eva, F. P., Birshtein, T. M., Gotlib, Yu. Ya.: Vysokomol. Soyed. *A 9*, 580 (1967)
344. Amiya, S. et al.: Polym. J. *6*, 194 (1974)
345. Mihailov, M. et al.: Macromolecules *6*, 194 (1973)
346. Miyamoto, T., Inagaki, H.: Macromolecules *2*, 554 (1969)
347. Koshevnik, A. Yu. et al.: Vysokomol. Soyed. *A 12*, 2103 (1970)
348. Krause, S., Roman, M.: J. Polym. Sci., Part A *3*, 1631 (1965)

349. Graham, R. K., Dunkelberger, D. L., Panchak, J. R.: J. Polym. Sci. *59*, 168 (1962)
350. Glusker, D. L., Galluccio, R. A., Evans, R. A.: J. Am. Chem. Soc. *86*, 187 (1964)
351. Miyamoto, T., Inagaki, H.: Polym. J. *1*, 46 (1970)
352. Inagaki, H., Miyamoto, T., Kamiyama, F.: Polym. Lett. *7*, 329 (1969)
353. Spévàček, J.: J. Polym. Sci., Polym. Phys. Ed. *16*, 523 (1978)
354. Miyamoto, T., Tomoshige, S., Inagaki, H.: Makromol. Chem. *176*, 3035 (1975)
355. Suzuki, H., Hiyoshi, T., Inagaki, H.: J. Polym. Sci., Symp. *61*, 291 (1977)
356. Tsvetkov, V. N., Boitsova, N. N.: Vysokomol. Soyed. *2*, 1176 (1960)
357. Kusakov, M. M. et al.: Vysokomol. Soyed. *B15*, 150 (1973)
358. Liu, K. J.: J. Polym. Sci., Part A-2 *5*, 1199 (1967)
359. Borchard, W., Pyrlik, M., Rehage, G.: Makromol. Chem. *145*, 449 (1971)
360. Takahashi, A., Ohwaki, S., Kagawa, I.: Bull. Chem. Soc., Japan *43*, 1262 (1970)
361. Schmidt, P. et al.: Europ. Polym. J. *11* 229 (1975)
362. Orlova, O. V. et al.: Dokl. Akad. Nauk SSSR *178*, 889 (1968)
363. Goode, W. E. et al.: J. Polym. Sci., *46*, 317 (1960)
364. Coleman, B. D., Fox, T. G.: J. Phys. Chem. *38*, 1065 (1963)
365. Coleman, B. D., Fox, T. G.: J. Polym. Sci., Part C *4*, 345 (1964)
366. Inagaki, H., Miyamoto, T., Kamiyama, F.: J. Polym. Sci., Part B *7*, 329 (1969)
367. Miyamoto, T., Tomoshige, S., Inagaki, H.: Polym. J. *6*, 564 (1974)
368. Spévàček, J., Schneider, B.: Colloid Polym. Sci. *258*, 621 (1980)
369. Mekenitskaya, L. I., Golova, L. K., Amerik, Yu. B.: Vysokomol. Soyed. *A22*, 893 (1980)
370. Ohara, K., Rehaga, G.: Colloid Polym. Sci. *259*, 318 (1981)
371. Bosscher, F., Keeksta, D., Challa, G.: Polymer *22*, 124 (1981)
372. Chapman, A. J., Billingham, N. C.: Europ. Polym. J. *16*, 21 (1980)
373. Hrouz, J. et al.: Makromol. Chem. *181*, 277 (1980)
374. Spévàček, J., Schneider, B.: Colloid Polym. Sci. *258*, 621 (1980)
375. Lohmeyer, J. H. G. M. et al.: J. Polym. Sci., Polym. Lett. Ed. *13*, 725 (1975)
376. Lohmeyer, J. H. G. M. et al.: Polymer *19*, 1171 (1978)
377. Schurer, J. W., De Boer, A., Challa, G.: Polymer *16*, 201 (1975)
378. Roerdink, E., Challa, G.: Polymer *21*, 509 (1980)
379. Spévàček, J., Schneider, B.: Makromol. Chem. *176*, 3409 (1975)
380. Andreyeva, G. A., Merkureva, A. V., Fedorova, L. A.: Vysokomol. Soyed. *A18*, 702 (1976)
381. Sulzberg, T., Cotter, R. J.: Macromolecules *1*, 554 (1968)
382. Mulvaney, J. E., Brand, R. A.: Macromolecules *13*, 244 (1980)
383. Geissler, U., Schulz, R. C.: Makromol. Chem. *181*, 1483 (1980)
384. Geissler, U., Schulz, R. C.: Makromol. Chem. *181*, 1495 (1980)
385. Turner, S. R.: Macromolecules *13*, 782 (1980)
386. Sulzberg, T., Cotter, R. J.: J. Polym. Sci., Part A-1 *8*, 747 (1970)
387. Bekturov, E. A., Bimendina, L. A.: Adv. Polym. Sci. *41*, 99 (1981)
388. Hatch, M. J., Dillon, J. A., Smith, H. B.: Ind. Eng. Chem. *44*, 1812 (1957)
389. Kossel, A.: J. Phys. Chem. *22*, 178 (1896)
390. Oosawa, F.: Biopolymers *6*, 1633 (1968)
391. Ise, N., Hosono, M.: J. Polym. Sci. *39*, 389 (1959)
392. Voorn, M. J.: Rec. Trav. Chim. *75*, 317 (1956)
393. Voorn, M. J.: Rec. Trav. Chim. *75*, 405 (1956)
394. Voorn, M. J.: Rec. Trav. Chim. *75*, 427 (1956)
395. Voorn, M. J.: Rec. Trav. Chim. *75*, 925 (1956)
396. Voorn, M. J.: Rec. Trav. Chim. *75*, 1021 (1956)
397. Michaeli, I., Overbeek, J. Th. G., Voorn, M. J.: J. Polym. Sci. *23*, 443 (1957)
398. Veis, A., Aranyi, C.: J. Phys. Chem. *64*, 1203 (1960)
399. Veis, A.: J. Phys. Chem. *65*, 1798 (1961)
400. Veis, A.: J. Phys. Chem. *67*, 1960 (1963)
401. Veis, A., Bodor, E., Mussell, S.: Biopolymers *5*, 37 (1967)

402. Tainaka, K., Yomosa, S.: J. Phys. Soc., Japan 48, 1791 (1980)
403. Tainaka, K.: Biopolymers 19, 1289 (1980)
404. Tainaka, K.: J. Phys. Soc., Japan 46, 1899 (1979)
405. Tang, Y. T.: Adv. Protein Chem. 16, 323 (1961)
406. Abe, K., Koide, M., Tsuchida, E.: J. Polym. Sci., Polym. Chem. Ed. 15, 2469 (1977)
407. Abe, K.: Dr. Thesis, Waseda University 1978
408. Abe, K., Koide, M., Tsuchida, E.: Polym. J. 9, 73 (1977)
409. Shinoda, K., Nakajima, A.: Bull. Inst. Chem. Res., Kyoto Univ. 53, 392 (1975)
410. Nakajima, A., Shinoda, K.: J. Colloid Interface Sci. 55, 126 (1976)
411. Shinoda, K., Nakajima, A.: Bull. Inst. Chem. Res., Kyoto Univ. 53, 400 (1975)
412. Mita, K., Ichimura, S., Zama, M.: Biopolymers 17, 2783 (1978)
413. Hammes, G. G., Schullery, S. E.: Biochemistry 7, 3882 (1968)
414. Nakajima, A. et al.: Polym. J. 7, 550 (1975)
415. Sato, H., Hayashi, T., Nakajima, A.: Polym. J. 8, 517 (1976)
416. Shinoda, K. et al.: Polym. J. 8, 208 (1976)
417. Gratzer, W. B., McPhie, P.: Biopolymers 4, 601 (1966)
418. Su Cho, C. et al.: Makromol. Chem. 180, 1951 (1979)
419. Komoto, T., Su Cho, C., Kawai, T.: Polym. Prep., Japan 28, 326 (1979)
420. Cundall, R. B. et al.: Polymer 20, 389 (1979)
421. Cundall, R. B., Lawton, J. B., Murray, D.: Makromol. Chem. 180, 2913 (1979)
422. Gelman, R. A., Blackwell, J.: Biopolymers 12, 541 (1973)
423. Gelman, R. A., Glaser, D. N., Blackwell, J.: Biopolymers 12, 1223 (1973)
424. Gelman, R. A., Blackwell, J.: Biopolymers 12, 1959 (1973)
425. Gelman, R. A., Blackwell, J.: Biopolymers 13, 139 (1974)
426. Shinoda, K., Hayashi, T., Nakajima, A.: Polym. J. 8, 216 (1976)
427. Tsuboi, M., Matsuo, K., Ts'o, P. O. P.: J. Mol. Biol. 15, 256 (1966)
428. Higuchi, S., Tsuboi, M.: Biopolymers 4, 837 (1966)
429. Matsuo, K., Tsuboi, M.: Bull. Chem. Soc., Japan 39, 347 (1966)
430. De Santis, P., Rizzo, R., Savino, M.: Macromolecules 6, 520 (1973)
431. Liu, H. J. et al.: Biopolymers 13, 1681 (1974)
432. Liu, H. J. et al.: Biopolymers 13, 649 (1974)
433. Blout, E. R., Idelson, M.: J. Am. Chem. Soc. 80, 4909 (1958)
434. Abe, K., Tsuchida, E.: Polym. J. 9, 79 (1977)
435. Baroqui, R., Tran, Q., Pefferkorn, E.: Macromolecules 12, 831 (1979)
436. Michaels, A. S.: U. S. Patent, 3,276,598 (1966)
437. U.S. Patent, 3,565,973 (1971)
438. U.S. Patent, 3,546,142 (1970)
439. U.S. Patent, 3,558,744 (1971)
440. Nakajima, A., Shinoda, K.: J. Appl. Polym. Sci. 21, 1249 (1977)
441. Hosono, M. et al.: J. Appl. Polym. Sci. 21, 2125 (1977)
442. Felsenfeld, G., Davies, D. R., Rich, A.: J. Am. Chem. Soc. 79, 2023 (1957)
443. Felsenfeld, G., Rich, A.: Biochim. Biophys. Acta 26, 457 (1957)
444. Hayes, F. N., Williams, D. L., Ratliff, R. L.: J. Am. Chem. Soc. 93, 4940 (1971)
445. Siano, D. E.: Biopolymers 17, 2897 (1978)
446. Tanaka, S., Baba, Y., Kagemoto, A.: Polym. J. 8, 325 (1976)
447. Ross, P. D., Scruggs, R. L.: Biopolymers 3, 491 (1965)
448. Suurkuusk, J. et al.: Biopolymers 16, 2641 (1977)
449. Rawitscher, M. A., Toss, P. D., Sturtevant, J. M.: J. Am. Chem. Soc. 85, 1915 (1963)
450. Stevens, C. L., Felsenfeld, G.: Biopolymers 2, 293 (1964)
451. Krakauer, H., Sturtevant, J. M.: Biopolymers 6, 491 (1968)
452. Baba, Y., Tanaka, S., Kagemoto, A.: Makromol. Chem. 178, 2117 (1977)
453. Baba, Y., Kagemoto, A.: Rep. Prog. Polym. Phys., Japan 21, 695 (1978)
454. Pochon, F., Michelson, A. M.: Proc. Natl. Acad. Sci., USA 53, 1425 (1965)
455. Springgate, M. W., Poland, D.: Biopolymers 12, 2241 (1973)

456. Davis, D. R., Rich, A.: J. Am. Chem. Soc. *80*, 1003 (1958)
457. Kallenbach, N., Drost, S.: Biopolymers *11*, 1613 (1972)
458. Tazawa, T., Tazawa, S., Ts'o, P. O. P.: J. Mol. Biol. *66*, 115 (1972)
459. Davis, D. R.: Nature *186*, 1030 (1960)
460. Baba, Y., Fujioka, K., Kagemoto, A.: Polym. J. *10*, 241 (1978)
461. Fujioka, K., Baba, Y., Kagemoto, A.: Rep. Prog. Polym. Phys., Japan *20*, 781 (1977)
462. Fujioka, K., Baba, Y., Kagemoto, A.: Rep. Prog. Polym. Phys., Japan *19*, 669 (1976)
463. Fujioka, K. et al.: Rep. Prog. Polym. Phys., Japan *22*, 743 (1979)
464. Takemoto, K.: J. Polym. Sci., Symp. *55*, 105 (1976)
465. Pitha, J.: Polymer *18*, 425 (1977)
466. Pitha, J., Pitha, P. M., Stuart, E.: Biochemistry *10*, 4595 (1971)
467. Overberger, C. G., Inaki, Y., Nambu, Y.: J. Polym. Sci., Polym. Chem. Ed., *17*, 1759 (1979)
468. Akashi, M. et al.: J. Polym. Sci., Polym. Chem. Ed. *17*, 905 (1979)
469. Hiraoka, K., Yokoyama, T.: Rep. Fac. Eng., Nagasaki Univ. *10*, 83 (1977)
470. Akashi, M. et al.: J. Polym. Sci., Polym. Chem. Ed. *17*, 747 (1979)
471. Tsuchida, E., Osada, Y., Ohno, H.: J. Macromol. Sci., Phys. *B 17*, 683 (1980)
472. Cho, C. S. et al.: Makromol. Chem. *179*, 1345 (1978)
473. Tsutsui, T., Tanaka, T.: Chem. Lett. *1976*, 1315 (1976)
474. Mori, T. et al.: J. Polym. Sci., Polym. Phys. Ed. *12*, 501 (1974)
475. Mori, T. et al.: Kobunshi Ronbunshu, *36*, 183 (1979)
476. Mori, T., Tanaka, R., Tanaka, T.: Polymer *18*, 1041 (1977)
477. Mori, T., Ogawa, K., Tanka, T.: J. Appl. Polym. Sci. *21*, 3381 (1977)
478. Mori, T. et al.: Kobunshi Ronbunshu *36*, 189 (1979)
479. Tanaka, T. et al.: J. Polym. Sci., Part A-1 *9*, 2745 (1971)
480. Kinoshita, M., Yamauchi, K., Imoto, M.: Prog. Polym. Sci., Japan *7*, 63 (1974)
481. Ohkubo, T., Ban, K., Ise, N.: Makromol. Chem. *175*, 49 (1974)
482. Uematsu, I., Honda, K.: Rep. Prog. Polym. Phys., Japan *8*, 111 (1965)
483. Osada, Y., Takeuchi, Y.: Polym. Prep., Japan *29*, 1389 (1980)
484. Kurokawa, Y., Ueno, K., Yui, N.: J. Colloid Interface Sci. *74*, 561 (1980)
485. Adelman, R. L., Klein, I. M.: J. Polym. Sci. *31*, 77 (1958)
486. Peterlin, A.: J. Polym. Sci. *12*, 45 (1954)
487. Fox, R. B.: Pure Appl. Chem. *34*, 235 (1973)
488. Gutmann, F., Lyons, J. E.: Organic Semiconductors, John Wiley & Sons, New York (1967)
489. Slifkin, M. A., Charge Transfer Interaction of Biomolecules, Academic Press, New York (1971)
490. Sulzberg, T., Cotter, R. J.: Macromolecules *2*, 146 (1969)
491. Sulzberg, T., Cotter, R. J.: Macromolecules *2*, 150 (1969)
492. Smetz, G., Balogh, V., Castille, Y.: J. Polym. Sci., Part C *4*, 1467 (1963)
493. Yang, N. C., Gaoni, Y.: J. Am. Chem. Soc. *86*, 5022 (1964)
494. Chang, D. M. et al.: J. Polym. Sci., Polym. Chem. Ed. *15*, 571 (1977)
495. Lumry, R., Rajender, S.: Biopolymers *9*, 1125 (1970)
496. Kabanov, V. A.: Macromol. Chem. (IUPAC Symposium) *8*, 121 (1973)
497. Stewart, G. T.: Mol. Cryst. *1*, 563 (1966)
498. Kabanov, V. A., Papisov, I. M.: Vysokomol. Soyed. *A 21*, 243 (1979)
499. Applequist, J., Dumle, V.: J. Am. Chem. Soc. *87*, 1450 (1965)
500. Dumle, V.: Biopolymers *9*, 353 (1970)
501. Lutsenko, V. V., Zezin, A. B., Lopatkin, A. A.: Vysokomol. Soyed. *A 16*, 2429 (1974)
502. Lutsenko, V. V., Zezin, A. B., Kalyuzhnaya, R. I.: Vysokomol. Soyed. *A 16*, 2411 (1974)
503. Latt, S. A., Sober, H. A.: Biochemistry *6*, 3293 (1964)
504. Papisov, I. M., Litmanovich, A. A.: Vysokomol. Soted. *A 19*, 716 (1977)
505. Baranovskii, V. Yu., Papisov, I. M.: Dokl. Akad. Nauk SSSR *217*, 123 (1974)
506. Anufrieva, Ye. V., et al.: Vysokomol. Soyed. *B 19*, 609 (1977)
507. Ohno, H., Tsuchida, E.: Makromol. Chem., Rapid Commun. *1*, 585 (1980)

508. Abe, K., Koide, M., Tsuchida, E.: Macromolecules *10*, 1259 (1977)
509. Anufrieva, Ye. V. et al.: Dokl. Akad. Nauk SSSR *232*, 1096 (1977)
510. Flory, P. J.: J. Polym. Sci. *49*, 105 (1961)
511. Beevers, R. B., White, E. H. T.: J. Polym. Sci., Part B *2*, 793 (1964)
512. Bach, D., Miller, I. R.: Biochim. Biophys. Acta *114*, 311 (1966)
513. Kim, H. J., Tonami, H: Kobunshi Kagaku *30*, 513 (1973)
514. Rogacheva, V. B., Zezin, A. B., Kargin, V. A.: Biofizika *15*, 389 (1970)
515. Kabanov, V. A. et al.: J. Polym. Sci., Part C *16*, 1079 (1967)
516. Zezin, A. B. et al.: Vysokomol. Soyed. *A 14*, 857 (1972)
517. Zezin, A. B., Kabanov, V. A., Kargin, V. A.: Biofizika *18*, 788 (1973)
518. Matsuda, H., Tsuchida, E., personal communication
519. Grutsh, J. F.: Environment. Sci. Technol. *12*, 1022 (1978)
520. Tsuchida, E., personal communication
521. Varma, A. J., Majewicz, T., Smid, J.: J. Polym. Sci., Polym. Chem. Ed. *17*, 1573 (1979)
522. Kabanov, N. M. et al.: Vysokomol. Soyed. *A 21*, 209 (1979)
523. Zezin, A. B. et al.: Vysokomol. Soyed. *A 19*, 118 (1977)
524. Kabanov, N. M. et al.: Vysokomol. Soyed. *A 21*, 1891 (1979)
525. Bere, A., Halene, C.: Biopolymers *18*, 2659 (1979)
526. Evreinova, T. N.: Concentration of Matter and Action of Enzymes on Coacervates, Nauka, Moscow (1966)
527. Oparine, A.: Dokl. Akad. Nauk SSSR *122*, 661 (1958)
528. Oparine, A.: Dokl. Akad. Nauk SSSR *104*, 581 (1955)
529. Evreinova, T. N.: Biofizika *29*, 1035 (1964)
530. Oparine, A.: Dokl. Akad. Nauk SSSR *148*, 943 (1963)
531. Ceperpovkaya, K. B.: Dokl. Akad. Nauk SSSR *135*, 1532 (1960)
532. Evreinova, T. N.: Dokl. Akad. Nauk SSSR *115*, 133 (1957)
533. Kuhn, W., Ramel, A., Waiters, D. H.: Angew. Chem. *70*, 314 (1958)
534. Katchalsky, A., Zwick, M.: J. Polym. Sci. *16*, 221 (1955)
535. Bolzer, E.: Arch. Biochem. Biophys. *73*, 144 (1958)
536. Hojo, N., Shirai, H., Mori, T.: Kogyo Kagaku Zasshi *74*, 273 (1971)
537. Kuhn, W.: Cazg. Chem. Ital. *92*, 951 (1962)
538. Kuhn, W., Ramel, A., Walters, D. H.: Chimia (Zurish) *12*, 123 (1958)
539. Mandelkern, L. et al.: J. Am. Chem. Soc. *84*, 383 (1962)
540. Mandelkern, L. et al.: Proc. Natl. Acad. Sci., USA *45*, 814 (1959)
541. Veen, G., Prins, W.: Nature, Phys. Sci. *230*, 70 (1971)
542. Osada, Y., Tsuchida, E.: J. Polym. Sci., Polym. Chem. Ed. *13*, 559 (1975)
543. Kargin, V. A., Kabanov, V. A., Kargina, O. V.: Dokl. Akad. Nauk SSSR *161*, 1131 (1965)
544. Kikuchi, Y., Kinoshita, M., Shima, K.: Makromol. Chem. *155*, 299 (1972)
545. Imanishi, Y. et al.: Biopolymers *11*, 181 (1973)
546. Osada, Y. et al.: Dokl. Akad. Nauk SSSR *191*, 399 (1970)
547. Golova, L. K., Amerik, Yu. B., Krentzel, B. A.: Vysokomol. Soyed. *B 12*, 8 (1970)
548. Golova, L. K., Amerik, Yu. B., Krentzel, B. A.: Vysokomol. Soyed. *B 12*, 565 (1970)
549. Kämmerer, H., Ozaki, Sh.: Makromol. Chem. *91*, 565 (1970)
550. Blumstein, A., Ponrathnam, S., Bellantoni, E.: ACS Polym. Prep. *20*, 640 (1979)
551. Blumstein, A., Ponrathnam, S., Bellantoni, E.: J. Polym. Sci., Polym. Lett. Ed. *18*, 299 (1980)
552. Ponrathnam, S., Blumstein, A., Bellantoni, E.: ACS Polym. Prep. *20*, 644 (1979)
553. Koetsier, D. W., Tan, Y. Y., Challa, G.: J. Polym. Sci., Polym. Chem. Ed. *18*, 1933 (1980)
554. Henrici-Olivé, G. and S.: Polym. Bull. *1*, 47 (1978)
555. Burillo, G., Chapiro, A., Mankowski, Z.: J. Polym. Sci., Polym. Chem. Ed. *18*, 327 (1980)
556. Ogata, N. et al.: J. Polym. Sci., Polym. Chem. Ed. *18*, 933 (1980)
557. Ogata, N. et al.: J. Polym. Sci., Polym. Chem. Ed. *18*, 939 (1980)
558. Shimidzu, T., Murakami, A., Konishi Y.: J. Chem. Soc., Perkin Trans. I *1979*, 20 (1979)
559. Hattori, M., Nakagawa, H., Kinoshita, M.: Makromol. Chem. *181*, 2325 (1980)

560. Akashi, M. et al.: J. Polym. Sci., Polym. Chem. Ed. *17*, 747 (1979)
561. Kataoka, S., Ando, T.: Kobunshi Ronbunshu *37*, 185 (1980)
562. Mason, R. G. et al.: Blood Compatibility of Some polymeric and Nonpolymeric Materials, *in vivo* and *ex vivo* Studies, Academic Press, New York (1969)
563. Halbert, S. P.: J. Biomed. Mat. Res. *4*, 549 (1970)
564. Petschek, H. E., Madras, P. N.: Thrombus Formation on Artificial Surfaces, Academic Press, New York (1969)
565. Kikuchi, Y., Hori, K.: Nippon Kagaku Kaishi *1980*, 1157 (1980)
566. Kikuchi, Y., Uemori, H.: Makromol. Chem. *176*, 821 (1975)
567. Fukuda, H., Kikuchi, Y.: Makromol. Chem. *178*, 2895 (1977)
568. Fukuda, H., Kikuchi, Y.: J. Biomed. Mat. Res. *12*, 531 (1978)
569. Rembaum, A.: J. Macromol. Sci., Chem. *A 4*, 715 (1970)
570. Jaques, L. B.: Pharmacol. Rev. *31*, 99 (1980)
571. Yen, S. P. S., Rembaum, A.: J. Biomed. Mat. Res. *1*, 83 (1971)
572. Carter R. L. et al.: Tumor Induction by Plastic Films, Academic Press, New York (1977)
573. Bruck, S. D.: J. Biomed. Mat. Res. *5*, 139 (1971)
574. Rotem, Z., Cox, R. A., Issacs, A.: Nature *197*, 564 (1963)
575. Field, A. K. et al.: Proc. Natl. Acad. Sci., USA *58*, 1004 (1967)
576. Mayhew, E. et al.: Molec. Pharmacol. *13*, 488 (1977)
577. Poste, G.: Int. Rev. Cytol. *33*, 157 (1972)
578. Poste, G., Pasternik, C. A.: Membrane Fusion, (Eds) G. Poste, G. L. Nicolson, p 305, North-Holland (1978)
579. Cohn, E. J. et al.: J. Am. Chem. Soc. *72*, 465 (1950)
580. Farr, R. S.: J. Infect. Dis. *103*, 239 (1958)
581. Polson, A. et al.: Biochim. Biophys. Acta *82*, 463 (1964)
582. Grambal, D.: Biochim. Biophys. Acta *251*, 54 (1971)
583. Richter, R. L., Morr, C. V., Reineccius, G. A.: J. Dairy Sci. *57*, 793 (1974)
584. Kamiya, T., Goto, K., Shinohara, I.: J. Polym. Sci., Polym. Chem. Ed. *17*, 561 (1979)
585. Riggopulos, P. N.: U.S. Patent, 3,549,016 (1970)
586. Fleming, S.: The Reverse Osmosis Performance of a Polyelectrolyte Complex Membrane, Academic Press, New York (1970)
587. Michaels, A. S.: U.S. Patent, 3,467,604 (1969)
588. Taylor, L. D.: U.S. Patent, 3,578,458 (1971)
589. Bixler, A.: U.S. Patent, 3,457,358 (1969)
590. Douglas, D. L.: U.S. Patent, 3,556,850 (1971)
591. Biddick, R. E.: Proceedings of the Sixth Annual Power Sources Symposium, Academic Press, New York (1968)
592. Tazuke, S., Nagahara, H.: Makromol. Chem. *181*, 2217 (1980)
593. Ohno, H., Abe, K., Tsuchida, E.: Makromol. Chem. *182*, 1253 (1981)
594. Ohno, H., Abe, K., Tsuchida, E.: Makromol. Chem. *182*, 1407 (1981)
595. Bungenberg de gong, H. G.: Trans. Far. Soc. *28*, 27 (1932)

Received February 17, 1981
H. J. Cantow (editor)

7 Subject Index

Author Index Volumes 1–45

Speciality Polymers

1981. 80 figures. V, 186 pages
(Advances in Polymer Science, Volume 41)
ISBN 3-540-10554-9

Contents/Information:

K. Takemoto, Y. Inaki: **Synthetic Nucleic Acid Analogs: Preparation and Interactions.** The functional monomers and polymers containing heterocyclic moieties have recently received much attention. Numerous studies have been devoted to the preparation and polymerization of these new monomeric species, which may find a number of application possibilities as polymeric drugs and other biomaterials. This article focusses on the authors work on specific base – base interactions between the nucleic acid analogues.
(85 references)

A. Y. Grosberg, A. R. Khokhlov: **Statistical Theory of Polymeric Lyotropic Liquid Crystals.** This article deals with following topics of the statistical physics of liquid-crystalline phase in the solutions of stiff chain macromolecules: problems of the phase diagram for the liquid-crystalline transition in the solutions of completely stiff macromolecules (rigid rods); conditions of formation of the liquid-crystalline phase in the solutions of semiflexible macromolecules; possibility of the intramolecular liquid-crystalline ordering in semiflexible macromolecules; structure of intramolecular liquid crystals and dependence of the properties of the liquid-crystalline phase on the microstructure of the polymer chain.
(45 references)

E. A. Bekturov, L. A. Binendina: **Interpolymer Complexes.** The problems of complex formation in different systems of interacting macromolecules, in polymer-polymer, polymer-alternating or statistical copolymer systems are discussed. The influence of solvent nature, the critical phenomena, equilibrium, selectivity and cooperativity in reactions are considered. The perspectives of development of this field of polymer science and the potential practical applications of interpolymer complexes are pointed out. (179 references)

N. Hagihara, K. Sonogashira, S. Takahashi: **Linear Polymers Containing Transition Metals in the Main Chain.** Recent studies on linear polymers containing transition metals in the main chain are reviewed on the basis of metallocene-containing polymers, linear "Werner Type" coordination polymers and polymers containing sigmabonded transition metals in the main chain. (56 references)

Springer-Verlag
Berlin
Heidelberg
New York

G. Govil, R. V. Hosur

Conformation of Biological Molecules

New Results from NMR

1982. 92 figures. VIII, 216 pages. (NMR, Volume 20)
ISBN 3-540-10769-X

Contents:

Recent developments in NMR have made it an indispensable tool in biochemistry and molecular biology. With the advent of FT-NMR techniques, it has become possible to solve problems of sensitivity, resolution and assignments for medium size molecules, and to study their conformational structure and dynamics in solutions.
Availability of labelled compounds is contributing to a wider use of NMR for biological problems. Applications in studies of multimolecular systems, dynamics of cellular chemistry, biological control and regulation and short lived reaction intermediates at enzyme active sites have started to appear in literature and major contributions of NMR to the field of molecular biology can be expected in the future. (1247 references).
This article reviews recent trends and developments in conformational studies on biological systems using NMR. The first two chapters deal with the theoretical principles and NMR techniques used in conformational analysis. The final four chapters deal with applications to different classes of biological molecules: nucleic acids and their components, amino acids, polypeptides and proteins, saccharides and polysaccharides and organisations in biomembranes. The emphasis is on basic principles and methodology in conformational analysis of biomolecules. Detailed coverage of the literature during the period from 1972 to 1980 is provided.

Springer-Verlag
Berlin
Heidelberg
NewYork